# SCIENCE

*Practical Applications*

Also in this series:

*(Practical Applications)*
Agriculture
Architecture
Art
Education
Medicine
Religion
Social and Political Science

*(Esoteric)*
Alchemy
Atlantis
Christian Rozenkreutz
The Druids
The Goddess
The Holy Grail

# RUDOLF STEINER

# SCIENCE

## An Introductory Reader

*Compiled with an introduction,
commentary and notes by
Howard Smith*

Sophia Books

All translations revised by Matthew Barton

Sophia Books
An imprint of Rudolf Steiner Press
Hillside House, The Square
Forest Row, RH18 5ES

www.rudolfsteinerpress.com

Published by Rudolf Steiner Press 2003

For earlier English publications of individual selections please
see pp. 233–5

The material by Rudolf Steiner was originally published in
German in various volumes of the 'GA' (*Rudolf Steiner
Gesamtausgabe* or Collected Works) by Rudolf Steiner Verlag,
Dornach. This authorized volume is published by permission of
the Rudolf Steiner Nachlassverwaltung, Dornach (for further
information see pp. 239–40)

This edition translated © Rudolf Steiner Press 2003

A catalogue record for this book is available from the British
Library

ISBN 1 85584 108 8

Cover photograph by Adam Hart-Davis, courtesy DHD Photo
Gallery. Cover design by Andrew Morgan
Typeset by DP Photosetting, Aylesbury, Bucks.
Printed and bound in Great Britain by Cromwell Press Limited,
Trowbridge, Wilts.

# Contents

# Introduction: Rudolf Steiner as Natural Scientist and Spiritual Scientist

## by Howard Smith

No one today can doubt that science makes an enormous impact on the lives of all of us, irrespective of whether we ourselves happen to be scientists or not. Our civilization is saturated by the technological applications of scientific discoveries, while scientific theories and ideas about the nature of the human being percolate to every level of society, informing government policy, economics, educational methods, medical ethics, etc. But what exactly do we mean by 'science'? How do we characterize this activity that is so much a part of the modern age? Where do the inner impulses come from which have helped create the world we now inhabit? Perhaps most importantly, to what extent does our current scientifically conditioned image of the human being and the cosmos reflect reality? Clearly, the answers to these questions concern us all.

This book, which offers a selection of texts from the published works of Rudolf Steiner, aims to throw some light on the above questions, at the same time serving as an introduction to his scientific thought. Between 1919 and 1923 Steiner gave several courses dealing directly with the nature of science and its various branches. However, practically every lecture course he gave, as well as his books and essays from as early as 1886, contain important references relevant to our study, irrespective of their particular theme. This selection is therefore an individual one, illustrative rather than comprehensive, and attempts to present the major landmarks of Steiner's many contributions to modern thought.

Steiner himself was well qualified to write and lecture on the science of his day. In 1879 (aged 18) he entered the Technische Hochschule (Polytechnic) in Vienna, where he studied biology, chemistry, physics and mathematics. Simultaneously he also immersed himself in philosophy and German literature, and at the age of 21 was invited to edit Goethe's scientific writings. Although Goethe was (and still is) regarded primarily as a poet and a playwright, the young Rudolf Steiner discovered in his scientific works a methodology which was to deeply influence his further work, and which he developed into a rigorous tool for natural and spiritual-scientific research.

Throughout his life Steiner showed a keen interest in all scientific developments, and often astounded colleagues with the breadth and depth of his knowledge of current research. Although science has made extremely rapid strides since Steiner's death in 1925, most of the groundwork for these advances had emerged during his lifetime. For example, knowledge of electricity was sufficiently advanced to facilitate radio transmission. The principles of physics and mechanics had advanced to the point of enabling motor cars to be mass-produced in America, and air flight was beginning. Chemists had already learned how to synthesize useful organic compounds (previously thought impossible), such as dyes, and the first synthetic polymer (bakelite) appeared in his lifetime. The young science of atomic physics was emerging, with the discovery of sub-atomic particles and three types of radiation, although at the time no one suspected what forces would eventually be unleashed. These researches and discoveries were all known by Steiner, but one can only speculate as to what extent his prophetic vision might have glimpsed later developments: nuclear power, genetic engineering, electronics, global computer networks, lasers, manned space-flight, synthetic medicines, and so forth.

Yet the value of Steiner's scientific insight does not depend on the extent to which specific facts at his disposal had already developed. He used his spiritual vision to shed light on science as a human activity, conditioned by our stage of evolutionary development, an activity in process of change from the earliest times, and on into the distant future. It is therefore only possible to evaluate Steiner's scientific thought with complete integrity when it is seen as part and parcel of his spiritual science (or anthroposophy). We are justified in using the term 'science' for investigating both the outer world of nature and the world of spirit when both worlds are approached with a certain quality of consciousness in which the mutual relationships of all phenomena are seen with the greatest possible clarity. Natural science and spiritual science are therefore two complementary aspects of human understanding, directed either to the 'natural' or the 'spiritual' world.

The relationship between the world of matter and the world of spirit is spoken of in many of Steiner's lectures and books. For example, in his book *Theosophy*, we find descriptions of how the kingdoms of nature have condensed or crystallized out of the primal spiritual world of archetypes and forces. Behind the mineral kingdom there is therefore a realm of forces and beings that provide the basic 'template' for the physical world. Indeed, behind all phenomena in nature there is a cosmic dimension which can be investigated, not by physical senses and instruments, but by senses adapted to the spiritual environment. Mathematics is often cited by Steiner as an example of an inner spiritual activity grasped in clear consciousness so as to become totally transparent. This mathematics is able to grasp the outer world since the substances and forces of this world originated in the same region in which the mathematics lives, i.e. the spiritual world.

> ... the condition of consciousness present in mathematical thinking is in fact what a person strives for who strives towards what I call imaginative knowledge. When we think mathematically, what is really the content of our soul? It is the numerical world, the spatial world, and so on... Thus we have in our soul the content of a particular field with a certain pictorial representation. To work in a similar condition of soul but

towards another pictorial content is what constitutes the development of imaginative cognition.[1]

Rudolf Steiner possessed the spiritual senses implied above (such as 'imaginative cognition'[2]), which lead to spiritual knowledge just as the physical senses lead to knowledge of the world of nature. It might therefore be assumed that 'belief' in spiritual science is necessary in order to understand Steiner's interpretation of natural science. This is not the case, however. He is at pains to explain that there is an unbroken line between the two realms. A secure grounding in natural science is a wonderful foundation for developing spiritual perceptions. This can happen when the bright spotlight of consciousness is turned away from physical phenomena and inwards towards the boundaries of ordinary scientific understanding. A contemplation of the apparent limits of knowledge can lead to the means to transcend those limits through developing greater capacities of perception and understanding. The spiritual activity of mathematics can lead to these higher capacities. Conversely, spiritual knowledge can throw light on physical phenomena by making visible the cosmic aspect, often with surprising results.

One problem sometimes encountered in reading Steiner, particularly in his descriptions of physical phenomena, is realizing what particular perspective he is talking from. Because he is constrained by the limitations of earthly language in conveying subtle spiritual concepts,

he tends to describe phenomena from a variety of per-
spectives. A physical object, when looked at with spiritual
senses, would appear devoid of all substance, and might
be described as pure thought or empty space or the like.
This is particularly evident in his descriptions of the atom,
which may seem incomprehensible unless the context is
understood.

In the selections chosen for this book, we shall see how
Steiner elucidates the wellsprings of the scientific impulse,
that is, the sources of scientific striving which arise from
spiritual necessity. What underlies the striving for
'objectivity'? What is the nature of scientific curiosity?
Why do we do experiments? Why do we try to understand
the world through mathematics? Steiner does not consider
these as self-evident or beyond question.

We shall also see how Steiner rescued 'Goetheanism'
from obscurity, how his lucid descriptions of Goethe's
method can help us towards a new science of the future.

We shall then consider Steiner's application and
extension of Goethe's approach to the realm where spiritual facts can be included in a scientific understanding of
the world. This includes selections from the 'Light Course'
(in Chapters 6 and 7) and from the groundbreaking
'Warmth Course' (in Chapters 8–11), which attempts to
build a framework for the states of matter and the different forces of nature.

The concluding chapter contains material in which
Steiner relates natural science to spiritual science from
several points of view. Although some of his comments

are critical, for example in surveying the effects of dubious scientific concepts on social conditions, we also find valuable indications that can help us towards a new orientation in natural science.

In working through these selections, the reader will notice marked variations in the character of different texts. This arises from the fact that some passages were much more carefully stenographed than others, whilst other texts were intrinsically very difficult to record. The Warmth Course in particular was a demonstration/lecture course for teachers, with Steiner setting up experiments, talking, and drawing on a blackboard, with frequent references to earlier sessions of the course. The style is therefore more 'chatty' than most of the other material, yet nonetheless breathtaking in its scope. We also have to reckon with the fact that some of the material is intrinsically more difficult to understand, and will need more study and rereading, whilst other sections are relatively straightforward.

As we proceed through the chapters it should become evident that Rudolf Steiner had enormous respect and admiration for what science had to offer. Yet he was all too aware of the dead-end that erroneous and incomplete interpretations of experimental results could lead to. A purely material science has led to a technology that has become dehumanized and fails to meet human needs for spiritual understanding. A true understanding of the natural world will some day enable natural science to blossom into a more spiritualized form, resulting in a technology which serves the true needs of human beings. Then natural science will contribute towards the same goal that Rudolf Steiner had set for spiritual science — to bring about a renewal of culture.

# 1. From Pre-science to Science

*It is generally believed that human consciousness itself is pretty much the same now as it was in ancient times, and that only its content has changed. Humans are believed to have evolved their ideas through their experience of the everyday world, from earlier states of superstition and ignorance to an increasingly sophisticated science which offers true insight. Rudolf Steiner, on the other hand, insists that the fabric of consciousness itself has changed dramatically over the millennia, and will continue to do so. A fundamental, purely inward development occurred in the fifteenth century, resulting in intensified intellectual activities in a recognizably modern sense. This laid the groundwork for the explosion of knowledge up to the present day, which appears to dwarf the achievements of previous long ages. Before this time, consciousness was significantly different, and the further back we go the more we find other faculties taking the place of science. Indeed, if we go far enough back, science becomes inconsequential. Here, Steiner traces the gradual emergence of the modern 'objective' mode of thinking, which confronts the world with questions, out of earlier forms in which there was far less distinction between object and subject.*

To sages of old, the universe was not the machine, the mechanical contraption that it is for human beings today when they look out into space. The cosmic spaces were like living beings, permeating everything with spirit and speaking to them in cosmic language. These sages experienced themselves within the spirit of world being. They felt how this cosmos in which they lived and moved spoke to them, how they could direct their questions about the riddles of the universe to the universe itself and how, out of the breadths of space, cosmic phenomena replied to them. This is how they experienced what we, in a weak and abstract way, call 'spirit' in our language. Spirit was experienced as the element that is everywhere and can be perceived from anywhere. Then human beings perceived things that even the Greeks no longer beheld with the eye of the soul, things that had faded into a nothingness for the Greeks.

This nothingness of the Greeks, which had been filled with living content for the earliest wise people of the post-Atlantean[3] age, was named in words customary for that time. Translated into our language, though weakened and abstract, those words would signify 'spirit'. What later became the unknown, the hidden God, was called spirit in those ages when he was known. This is the first thing to know about those ancient times.

The second thing to know is that when a human being looked into himself with his soul and spirit vision, he beheld his soul. He experienced it as originating from the spirit that later on became the unknown God. The experience of the ancient sage was such that he designated the human soul by a term that would translate in

our language into 'spirit mes̲
senger'.

If we put into a diagram what was
earliest times, we can say: The spirit ̲
world-embracing element, apart from ̲
nothing, and by which everything was p̲
spirit, directly perceptible in its archetypa̲          was
sought and found in the human soul, inasm ̲n as the
latter recognized itself as the messenger of this spirit. Thus
the soul was referred to as the 'messenger'.

A third aspect was external nature with all that today is
called the world of physical matter, of corporeality. I said
above that apart from spirit there was nothing, because
spirit was perceived by direct vision everywhere in its
archetypal form. It was seen in the soul, which realized the
spirit's message in its own life. But the spirit was likewise
perceived in what we now call nature, the world of corporeal things. Even this bodily world was looked upon as
an image of the spirit.

Spirit: archetypal form
Soul: messenger
Body: image

In those ancient times, people did not have the conceptions that we have today of the physical world. Wherever
they looked, at whatever thing or form of nature, they
beheld an image of the spirit, because they were still
capable of seeing the spirit. The image nearest to man was
the human body, a fragment of nature. Inasmuch as all

phenomena of nature were images of the spirit, the body of man too was an image of the spirit. So when this ancient human looked at himself, he recognized himself as a threefold being. In the first place, the spirit lived in him as in one of its many mansions. Man knew himself as spirit. Secondly, man experienced himself within the world as a messenger of this spirit, hence as a soul being. Thirdly, man experienced his corporeality; and by means of this body he felt himself to be an image of the spirit. Hence, when man looked upon his own being, he perceived himself as a threefold entity of spirit, soul and body: as spirit in his archetypal form; as soul, the messenger of God; as body, the image of the spirit.

This ancient wisdom contained no contradiction between body and soul or between nature and spirit, because people knew that spirit was within man in its archetypal form. The soul was none other than the message transmitted by spirit; the body was image of spirit. Likewise, no contrast was felt between man and surrounding nature because one bore an image of spirit in one's own body, and the same was true of each body in outer nature. Hence an inner kinship was experienced between one's own body and those in outer nature, and nature was not felt to be different from oneself. Man felt himself at one with the whole world. He could feel this because he could behold the archetype of spirit and because the cosmic expanses spoke to him. In consequence of the universe speaking to man, science simply could not exist. Just as we today cannot build a science of external nature out of what lives in our memory, ancient man could not develop one because, whether he looked into himself or outwards at nature, he beheld the same image of spirit. No contrast existed between man himself and nature, and there was none between soul and body. The correspondence of soul and body was such that, in a manner of speaking, the body was only the vessel, the artistic reproduction, of the spiritual archetype, while the soul was the mediating messenger between the two. Everything was in a state of intimate union. There could be no question of comprehending anything. We grasp and comprehend what is outside our own life. Anything that we carry within ourselves is directly experienced and need not be first comprehended.

Prior to Roman and Greek times, this wisdom born of direct perception still lived in the mysteries. The pupils learned them from the teachers, but the teachers could no longer see them, at least not in the vividness of ancient times. Indeed, in former times souls had the inner resilience needed to say to themselves: In the inward perception of the spirit indwelling me, I myself am something divine. But then it gradually happened that, for direct perception, the spirit no longer inhabited the soul. No longer did the soul experience itself as the spirit's messenger, for to be a messenger one must fully know the one from whom the message comes. Now the soul only felt itself as the bearer of the Logos, the spirit image, though this spirit image was vivid in the soul. This expressed itself in love for this God whose image still lived in the soul. But the soul no longer felt like the messenger, only the carrier, of an image of the divine spirit. One can say that a different form of knowledge arose when man looked into his inner being. The soul declined from messenger to bearer.

Soul: bearer
Body: force

Since the living spirit had been lost to human perception, the body no longer appeared as the image of spirit. To recognize it as such an image, one would have had to perceive the archetype. Therefore, for this later age, the body changed into something that I would like to call 'force'. The concept of force emerged. The body was pictured as a complex of forces, no longer as a reproduction, an image, that bore within itself the essence of what it reproduced. The human body became a force which no longer bore the substance of the source from which it originated.

Not only the human body, but in all of nature, too, forces had to be pictured everywhere. Whereas formerly nature in all its aspects had been an image of spirit, now it had become forces flowing out of spirit. This, however, implied that nature began to be something more or less foreign to man. One could say that the soul had lost something since it no longer contained direct spirit awareness. Speaking crudely, I would have to say that the soul had inwardly become more tenuous, while the body, the external corporeal world, had gained in robustness.

Earlier, as an image, it still possessed some resemblance to the spirit. Now it became permeated by the element of force. The complex of forces is more robust than the image in which the spiritual element is still recognizable. Hence, again speaking crudely, the corporeal world became denser while the soul became more tenuous. Now, a contrast that had not existed before arose between the soul, grown more tenuous, and the increased density of the corporeal world. Previously, the unity of spirit had been perceived in all things. Now, there arose the contrast between body and soul, man and nature. Man now felt himself divided as well from nature, something that also had not been the case in the ancient times. Human beings now struggle to comprehend the connection between, on the one hand, the soul, that lacks spirit reality, and, on the other hand, the body that has become dense, has turned into force, into a complex of forces.

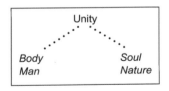

And human beings struggle to feel and experience the relationship between man and nature. But everywhere nature is force. At that time no conception at all existed as yet of what we call today 'the laws of nature'. People did not think in terms of natural laws; everywhere and in everything they felt the forces of nature. When a person looked into his own being, he did not experience a soul that—as was the case later on—bore within itself a dim will, an almost equally dim feeling, and an abstract thinking. Instead, he experienced the soul as bearer of the living Logos, something that was not abstract and dead, but a divine living image of God.

There came a time when even awareness of the Logos indwelling the soul was lost. Only then, at that point, did the modern era of thinking begin. The soul now no longer contains the living Logos. Instead, when it looks into itself, it finds ideas and concepts, which finally lead to abstractions. The soul has become even more tenuous. A third phase begins. Once upon a time, in the first phase, the soul experienced the spirit's archetype within itself. It saw itself as the messenger of spirit. In the second phase, the soul inwardly experienced the living image of God in the Logos; it became the bearer of the Logos.

Now, in the third phase, the soul becomes, as it were, a vessel for ideas and concepts. These may have the certainty of mathematics, but they are only ideas and concepts. The soul experiences itself at its most tenuous, if I may put it so. Again the corporeal world increases in robustness. This is the third way in which man experiences himself. He cannot as yet give up his soul element completely, but he experiences it as the vessel for the realm of ideas. He experiences his body, on the other hand, not only as a force but as a spatial body.

Soul: realm of ideas
Body: spatial corporeality

The body has become still more robust. Man now denies the spirit altogether. Here we come to the 'body' that Hobbes, Bacon, and Locke spoke of. Here, we meet 'body' at its densest. The soul no longer feels a kinship to it, only an abstract connection that degenerates over the course of time.

In place of the earlier concrete contrast of soul and body, man and nature, another contrast arises that leads further and further into abstraction. The soul that formerly appeared to itself as something concrete — because it experienced in itself the Logos-image of the divine — gradually transforms itself to a mere vessel of ideas. Whereas before, in the ancient spiritual age, it had felt akin to everything, it now sees itself as subject and regards everything else as object, feeling no further kinship with anything.

The earlier contrast of soul and body, man and nature, increasingly became the merely theoretical epistemological contrast between the subject that is within a person and the object without. Nature changed into the object of knowledge. It is not surprising that out of its own needs knowledge henceforth strove for the 'purely objective'.

But what is this purely objective aspect? It is no longer what nature was to the Greeks. The objective is external corporeality in which no spirit is any longer perceived. It is nature devoid of spirit, to be comprehended from without by the subject.

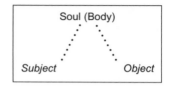

The striving to develop science must therefore be pictured
as emerging from earlier faculties of mankind. A time had
to come when this science would appear. It had to develop
the way it did. We can follow this if we focus clearly on the
three phases of development that I have just described.

We see how the first phase extends to the eighth century
BC. The second extends from then to Nicholas Cusanus.[4]
We find ourselves in the third phase now. The first is
pneumatological, directed to the spirit in its primeval
form. The second is mystical, in the broadest possible
sense. The third is mathematical. Considering these significant characteristics, therefore, we trace the first
phase — ancient pneumatology — as far as the ancient wise
men. Magical mysticism extends from there to Meister
Eckhart and Nicholas Cusanus. The age of mathematics-
based natural science proceeds from Cusanus into our
own time and continues further.

## 2. The Origin of Mathematics

*Mathematics is regarded as the 'gold standard' of clear think-
ing, and mathematical models are to be found everywhere.
Indeed, the more any field of knowledge can be quantified and
recast in mathematical form, so much the better for the modern
scientist. As the philosopher Kant expressed it: 'In every individual science there is only so much real knowledge as there is
mathematics.' But why do we feel so secure with mathematical
explanations? Curiously, some scientists regard mathematics
as merely an arbitrary human invention which happens to be
very useful. However, Rudolf Steiner shows the objective roots
of mathematical activity by tracing it to processes within the
human being. It arises in the course of evolution as a metamorphosis of inner bodily processes, which were once experienced but are now externalized and projected into the world.
Grasping the world mathematically gives us confidence
because we are linking the world with inwardly felt processes.
At the same time, understanding this mathematical activity for
what it is helps us to see that there are indeed other ways of
comprehending the world. The gradual externalization of
mathematics is exemplified by Ptolemy (b. c. AD 75), whose
geocentric astronomy was based on inwardly felt mathematical
relationships, Copernicus (1473–1543), who saw the planets as
independent of the earth yet retained some connection with the
mathematics in his feeling, and Newton (1642–1727), who*

*was able to view mathematics as totally disconnected from man.*

If the character of scientific thinking is to be correctly understood, it must be through the special way in which man relates to mathematics and mathematics relates to reality. Mathematics had gradually become what I would term a self-sufficient inward capacity for thinking. What do I mean by that?

The mathematics existing in the age of Descartes and Copernicus can certainly be described more or less in the same terms as apply today. Take a modern mathematician, for example, who teaches geometry, and who uses his analytical formulas and geometrical concepts in order to comprehend some physical process. As a geometrician, this mathematician starts from the concepts of Euclidean geometry, three-dimensional space (or merely dimensional space, if he thinks of non-Euclidean geometry). In three-dimensional space he distinguishes three mutually perpendicular directions that are otherwise identical. Space, I would say, is a self-sufficient form that is simply placed before one's consciousness in the manner described above without questions being raised such as: Where does this form come from? Or: Where do we get our whole geometrical system?

In view of the increasing superficiality of psychological thinking, it was only natural that man could no longer penetrate to those inner depths of soul where geometrical

thought has its foundations. Man takes his ordinary consciousness for granted and fills this consciousness with
mathematics that has been thought out but not experienced. As an example of what is thought out but not
experienced, let us consider the three perpendicular
dimensions of Euclidean space. Man would have never
thought of these if he had not experienced a threefold
orientation within himself. One orientation that man
experiences in himself is from front to back. We need
only recall how, from the modern, external anatomical
and physiological point of view, intake and excretion of
food, as well as other processes in the human organism,
take place from front to back. The orientation of these
specific processes differs from the one that prevails
when, for example, I do something with my right arm
and make a corresponding move with my left arm. Here,
the processes are oriented to left and right. Finally, in
regard to the last orientation, man grows into it during
earthly life. In the beginning he crawls on all fours and
only gradually stands upright, so that this last orientation
flows within him from above downwards and up from
below.

As matters stand today, these three orientations in the
human being are regarded very superficially. These processes—front to back, right to left or left to right, and
above to below—are not inwardly experienced so much
as viewed externally. If it were possible to go back into
earlier ages with true psychological insight, one would
perceive that these three orientations were inward

experiences for the people of former times. Today our thoughts and feelings are still half-way acknowledged as inward experiences, but the person of a bygone age had a real inner experience, for example, of the front-to-back orientation. He had not yet lost awareness of the decrease in intensity of taste sensations from front to back in the oral cavity. The qualitative experience that taste was strong on the tip of the tongue, then grew fainter and fainter as it receded from front to back, until it disappeared entirely, was once a real and concrete experience. The orientation from front to back was felt in such qualitative experiences. Our inner life is no longer as intense as it once was. Therefore, today, we no longer have experiences such as this. Likewise man today no longer has a vivid feeling for the alignment of his axis of vision in order to focus on a given point by shifting the right axis over the left. Nor does he have a full concrete awareness of what happens when, in the right-left orientation, he relates his right arm and hand to the left arm and hand. Even less does he have a feeling that would enable him to say: A thought illuminates my head and, moving in the direction from above to below, it strikes into my heart. Such a feeling, such an experience, has been lost to man along with the loss of all inwardness of world experience. But it did once exist. Man did once experience the three perpendicular orientations of space within himself. And these three spatial orientations — right-left, front-back, and above-below — are the basis of the three-dimensional framework of space, which is only an abstraction of the immediate inner experience described above.

So what can we say when we look back at the geometry of earlier times? We can put it like this. In those days it was obvious to people that merely through being human geometrical elements revealed themselves in their own life. By extending their own above-below, right-left, and front-back orientations, they could grasp the world out of their own being.

Try to sense the tremendous difference between this mathematical feeling bound to human experience, and the bare, bleak mathematical layout of analytical geometry, which establishes a point somewhere in abstract space, draws three coordinating axes at right angles to each other, and thus isolates this thought-out space scheme from all living experience. But man has in fact torn this thought-out spatial diagram out of his own inner life. So if we are to understand the origin of the later mathematical way of thinking that science adopted, if we are to correctly comprehend its self-sufficient presentation of structures, we must trace it back to the mathematics founded on human experience of a bygone age. Mathematics in former times was something completely different. What was once present in a sort of dreamlike experience of three-dimensionality and then grew abstract has become completely unconscious today. As a matter of fact, we still produce mathematics from our own three-dimensional experience. But the way in which we derive this spatial schema from our experiences of inward orientation is completely unconscious. None of this rises into consciousness except the finished spatial diagram. The same is true of all completed mathematical structures. They have all been severed from their roots. I chose the example of the spatial schema, but I could just as well mention any other mathematical category taken from algebra or arithmetic. They are nothing but schemata drawn from immediate human experience and raised into abstraction.

Where is the human factor if I imagine an abstract point somewhere in space, crossed by three perpendicular directions, and then apply this scheme to a process perceived in actual space? It is completely divorced from man, something quite inhuman. This non-human element, which has appeared in recent times in mathematical thinking, was once human. But when was it human?

It was human when man did not only experience in his movements and his inward orientation in space that he stepped forward from behind and moved in such a way that he was aware of his vertical as well as the horizontal direction, but when he also felt the blood's inward activity in all such movement, in all such inner geometry. There is always blood activity when I move forward. Think of the blood activity present when, as an infant, I lifted myself up from the horizontal to an upright position! Behind man's movements, behind his experience of the world by virtue of movement (which can also be, and at one time was, an inward experience) there stands the experience of the blood. In fact, if you run in a triangle, you can have a particular inner experience of the blood. You will have a different one if you run in a square.

What is outwardly quantitative and geometric is inwardly intensely qualitative in our experience of the blood.

It is surprising, very surprising, to discover that ancient mathematics spoke quite differently about the triangle and square. Modern nebulous mystics describe great mysteries, but there is no great mystery here. It is only what a person would have experienced inwardly in the blood when he walked the outline of a triangle or a square, not to mention the blood experience corresponding to the pentagram. In the blood the whole of geometry becomes qualitative inward experience. We arrive back at a time when one could truly say, as Mephistopheles does in Goethe's *Faust*, 'Blood is a very special fluid.' This is because, inwardly experienced, the blood absorbs all geometrical forms and makes of them intense inner experiences. Thereby man learns to know himself as well. He learns to know what it means to experience a triangle, a square, a pentagram; he becomes acquainted with the projection of geometry on the blood and its experiences. This was once mysticism. Not only was mathematics closely related to mysticism, it was in fact the external side of movement, of the limbs, while the inward side was the blood experience. For the mystic of bygone times all of mathematics transformed itself out of a sum of spatial formations into what is experienced in the blood, into an intensely mystical rhythmic inner experience.

We can say that once upon a time man possessed a knowledge that he experienced, that he was an integral part of, and that at the point in time that I have mentioned, he lost this oneness of self with the world, this participation in the cosmic mysteries. He tore mathematics loose from his inner being. No longer did he have the experience of movement; instead, he mathematically constructed the relationships of external movement. He no longer had the blood experience; the blood and its rhythm became something quite foreign to him. Imagine what this means. The human being tears mathematics free from his body and it becomes something abstract. He loses his understanding of the blood experience. Mathematics no longer goes inwards. Picture this as a soul mood that arose at a specific time. Previously the soul had a different mood than it did later on. Formerly it sought the connection between blood experience and experience of movement; later, it completely separated them. It no longer related the mathematical and geometrical experience to its own movement. It lost the blood experience. Think of this as real history, as something that occurs in the changing moods of evolution. Truly, someone who lived at the earlier age, when mathematics was still mysticism, put his whole soul into the universe. He measured the cosmos against himself. He lived in astronomy.

Modern man applies his system of coordinates to the universe but keeps himself out of it. Earlier, man sensed a blood experience with each geometrical figure. Modern man feels no blood experience; he loses the relationship to his own heart, where the blood experiences are centred. Is it imaginable that in the seventh or eighth century, when the soul still felt movement as a mathematical experience and blood as a mystical experience, anybody would have founded a Copernican astronomy with a system of coordinates simply imposed on the universe and totally divorced from man? No, this became possible only when a specific soul constitution arose in evolution. And after that something else became possible as well. Inward blood awareness was lost. Now the time had come to discover the movements of the blood externally through physiology and anatomy. Hence you have this change in evolution: on the one hand Copernican astronomy, on the other the discovery of the circulation of the blood by Harvey, a contemporary of Bacon and Hobbes. A world view gained by abstract mathematics cannot produce anything like the ancient Ptolemaic theory, which was essentially bound up with man and the living mathematics he experienced within himself. Now, instead, people experience an abstract system of coordinates starting with an arbitrary zero point. No longer do we have the inward blood experience; instead, we discover the physical circulation of the blood with the heart at the centre.

The birth of science thus placed itself into the whole context of evolution in both its conscious and unconscious processes. Only in this way, out of the truly human element, can one understand what actually happened, what had to happen in recent times for science — so self-evident today — to come into being in the first place. Only thus could it even occur to anybody to conduct such investigations as led, for example, to Harvey's discovery of the circulation of the blood.

## Ptolemy, Copernicus and Newton: the development of astronomy

Think back to the other world view that I have previously described to you, when all corporeality was regarded as image of the spirit. Looking at a body one found in it the image of spirit. One then looked back on oneself, on what — in union with one's own divine nature — one experienced as mathematics through one's own bodily constitution. Just as a work of art is not something obscure but is recognized as the image of the artist's ideas, so one found in corporeal nature the mathematical images of what one had experienced with one's own divine nature. The bodies of external nature were images of the divine spiritual. The instant that mathematics is separated from man and is regarded only as an attribute of bodies that are no longer seen as a reflection of spirit, agnosticism creeps into knowledge.

Take a concrete example, the first phenomenon that confronts us after the birth of scientific thinking, the Copernican system. It is not my intention today or in any of these lectures to defend either the Ptolemaic or the Copernican system. I am not advocating either one. I am only speaking of the historical fact that the Copernican system has replaced the Ptolemaic. What I say today does not imply that I favour the old Ptolemaic system over the Copernican. But this must be said as a matter of history. Imagine yourself back in the age when man truly experienced his own orientation in space: above-below, right-left, front-back. He could experience this only in connection with the earth. He could, for example, experience the vertical orientation in himself only in relation to the direction of gravity. He experienced the other two in connection with the four compass points according to which the earth itself is oriented. All this he experienced together with the earth as he felt himself standing firmly on it. He thought of himself not just as a being that begins with the head and ends at the soles of the feet. Rather, he felt himself penetrated by the force of gravity, which had something to do with his being but did not cease at the soles of his feet. Hence, feeling himself within the nature of the gravitational force, man felt himself one with the earth. In his concrete experience, the starting-point of his cosmology was thus given by the earth. Therefore he felt the Ptolemaic system to be justified.

Only when man severed his own being from mathematics, only then was it possible also to sever mathematics from the earth and to found an astronomical system with its centre in the sun. Man had to lose the old inner self-experience before he could accept a system with its centre outside the earth. The rise of the Copernican system is therefore intimately bound up with the transformation of civilized mankind's soul mood. The origin of modern scientific thinking cannot be separated from the general mental and soul condition, but must be viewed in context with it.

Copernicus tried to grasp astronomy with abstract mathematical ideas. On the other hand, Newton shows mathematics completely on its own. Here I do not mean single mathematical deductions, but mathematical thinking in general, entirely divorced from human experience. This sounds somewhat radical and objections could certainly be made to what I am thus describing in broad outlines, but this does not alter the essential facts. Newton is pretty much the first to approach the phenomena of nature with abstract mathematical thinking. Hence, as a kind of successor to Copernicus, Newton becomes the real founder of modern scientific thinking.

It is interesting to see in Newton's time and in the age that followed how civilized humanity is at pains to come to terms with the immense transformation in soul configuration that occurred as the old mathematical-mystical view gave way to the new mathematical-scientific style. The thinkers of the time find it difficult to come to terms with this revolutionary change. It becomes all the more evident when we look into the details, the specific problems with which some of these people wrestled. See how Newton, for instance, presents his system by trying to relate it to the mathematics that has been severed from man. We find that he postulates time, place, space and motion. He says in effect in his *Principia*: I need not define place, time, space and motion because everybody understands them. Everybody knows what time is, what space, place and motion are, hence these concepts, taken from common experience, can be used in my mathematical explanation of the universe. People are not always fully conscious of what they say. In life, it actually happens seldom that a person fully penetrates everything he says with his consciousness. This is true even of the greatest thinkers. Thus Newton really does not know why he takes place, time, space and motion as his starting-points and feels no need to explain or define them, whereas in all subsequent deductions he is at pains to explain and define everything. Why does he do this? The reason is that in regard to place, time, motion and space all cleverness and thinking avail us nothing. No matter how much we think about these concepts, we grow no wiser than we were to begin with. Their nature is such that we experience them simply through our common human nature and must take them as they come. Therefore one can say that Newton takes the trivial idea of space just as he finds it, but then he begins to calculate. But, due to the particular quality of thinking in his age, he already has an abstracted mathematics and geometry, and therefore he penetrates spatial phenomena and processes of nature with abstract mathematics. Thereby he sunders natural phenomena from man. In fact, in Newton's physics we meet for the first time ideas of nature that have been completely divorced from man. Nowhere in earlier times were conceptions of nature so torn away from man as they are in Newtonian physics.

# 3. The Roots of Physics and Chemistry, and the Urge to Experiment

*Over the portico of the Temple of Apollo at Delphi are carved the words 'Know Thyself', which is often taken to mean 'Man, know thyself and thou shalt know the World'. The ancients regarded man as a microcosm which reflected the universal macrocosm. The history of scientific discovery is nothing more than the gradual transformation of inner experiences into consciousness, where they are grasped as external forms. The mechanical forces in the organism externalize as physics, while the powerful chemical forces of transformation externalize as chemistry.*

*The fact that organisms contain great wisdom is being exploited in a practical way in some university research departments. For example, in modern bio-engineering, the principle of 'reverse engineering' is used. This seeks to analyse biological systems (regarded as being 'engineered by nature') with the aim of discovering solutions to engineering problems such as making new components and building networks.*

*When something inward is expressed in outer form, the soul needs to find a new relationship to it. In ancient times this resulted in rituals, which today have become metamorphosed into a different type of practical activity – the experiment.*

We can show in many ways that in older times there was no feeling that anything was completely divorced from man. Within himself, man experienced the processes and events as they occurred in nature. When he observed the fall of a stone, for example, in external nature (an event physically separated from him) he experienced the essence of movement. He experienced this by comparing it with what such a movement would be like in himself. When he saw a falling stone, he experienced something like this: 'If I wanted to move in the same way, I would have to acquire a certain speed, and in a falling stone the speed differs from what I observe, for instance, in a slowly crawling creature.' He experienced the speed of the falling stone by applying his experience of movement to the observation of the falling stone. The processes of the external world that we study in physics today were in fact also viewed objectively by people of former times, but they gained their knowledge with the aid of their own experiences in order to rediscover in the external world the processes going on within themselves.

Until the beginning of the fifteenth century, all the conceptions of physics were pervaded by something of which one can say that it brought even the physical activities of objects close to the inner life of man. Man experienced them in unison with nature. But with the onset of the fifteenth century begins the divorce of the observation of such processes from man. Along with it came the severance of mathematics, a way of thinking which from then on was combined with all science. Inner experience in the physical body was totally lost. What can be termed the inner physics of the human being was lost. External physics was divorced from man, along with mathematics. The progress thereby achieved consisted in the objectifying of the physical. What is physical can be looked at in two ways. Staying with the example of the falling stone, it can be traced with external vision. It can also be brought together with the experience of the speed that would have to be achieved if one wanted to run as fast as the stone falls. This produces comprehension that goes through the whole man, not one related only to visual perception.

To see what happened to the older world view at the dawn of the fifteenth century, let us look at a man in whom the transition can be observed particularly well — Galileo (1564–1642). Galileo, one can say, discovered the laws governing falling objects. Galileo's main aim was to determine the distance travelled in the first second by a falling body. The older world view placed visual observation of the falling stone side by side with inward experience of the speed needed to run at an equal pace. Inner experience was placed alongside that of the falling stone. Galileo also observed the falling stone, but he did not compare it with inward experience. Instead, he measured the distance travelled by the stone in the first second of its fall. Since the stone falls with increasing speed, Galileo also measured the following segments of its path. He did not align this with any inward experience, but with an externally measured process that had nothing to do with man, a process that was completely divorced from man. Thus, in perception and knowledge, the physical was so completely removed from man that he was not aware that he had the physical within him as well.

From the fifteenth century onwards, the whole orientation of the human mind was led to such a point that we can fairly say that man forgot his own inward experience. This happens first with the inner experience of the physical organism — man forgets it. What Galileo thought out and applied to matters close to man, such as the law of inertia, was now applied over a wide context. And it was indeed merely thought out in an abstract way, even if Galileo was dealing with things that can be observed in nature.

Through Newton the whole abstracted physical mode of conception becomes generalized to such an extent that it applies to the whole universe. In short, the aim is to completely forget all experience within man's physical body; to objectify what was formerly pictured as closely related to the experience of the physical body; to view it in outer space independent of physical corporeality, after first tearing it out of a body-based experience; and to find ways to speak of space without even thinking about the human being. Through separation from the physical body, through separation of nature's phenomena from man's experience in the physical body, modern physics arises. It comes into existence along with this separation of certain processes of nature from self-experience within the physical human body. Self-experience is forgotten.

By permeating all external phenomena with abstract mathematics, this kind of physics could no longer understand man. What had been separated from man could not be reconnected. In short, there emerges a total inability to bring science back to man.

In physical respects you do not notice this quite so much; but you do notice it if you ask: What about man's self-experience in the etheric body[5] in this subtle organism? Man experiences a good deal in it. But this was separated from man even earlier and more radically. This abstraction, however, was not as successful as in physics. Let us go back to a scientist of the first Christian centuries, the physician Galen.[6] Looking at what lived in external nature and following the traditions of his time, Galen distinguished four elements — earth, water, air and fire (we would say heat or warmth). We see these if we look at nature. But, looking inward and focusing on the self-experience of the etheric body, one asks: How do I experience these elements, the solid, the watery, the airy and the fiery, in myself? In those times the answer was: I experience them with my etheric body. One experienced it as inwardly felt movements of the fluids: the earth as 'black gall', what was watery as 'phlegm', the airy as 'pneuma'[7] (everything taken in through the breathing process), and warmth as 'blood'. In the fluids, in what circulates in the human organism, the same thing was experienced as what was observed externally. Just as the movement of the falling stone was accompanied by an experience in the physical body, so the elements were experienced in inward processes. The metabolic process, where (so it was thought) gall, phlegm and blood work into each other, was felt as the inner experience of one's own body, but a form of inward experience to which corresponded the external processes occurring between air, water, fire and earth.

Warmth — Blood: ego organization
Air — Pneuma: astral body
Water — Phlegm: etheric body — chemistry
Earth — Black Gall: physical body — physics

Here, however, we did not succeed in completely forgetting all inner life and still satisfying external observation. In the case of a falling body, one could measure something: for example, the distance travelled in the first second. One arrived at a 'law of inertia' by thinking of moving points that do not alter their condition of movement but maintain their speed. By attempting to dispel from inner experience something that the ancients strongly felt to belong specifically to it — that is, the four elements — one was able to forget the inner content, yet could not find any feasible measuring system in the external world. Therefore the attempt to objectify what related to these matters, as was done in physics, has remained basically unsuccessful to this day. Chemistry could have become a science that ranked alongside physics, if it had been possible to externalize the etheric body to the extent accomplished with the physical body. In chemistry, however, unlike physics, we speak to this day of something rather undefined and vague, when referring to its laws.[8] What was done by physics in regard to the physical body was actually also the aim of chemistry in regard to the etheric body. Chemistry states that if substances combine chemically, and in doing so can completely alter their properties, a natural change is taking place.

But if one wants to go beyond this conception, which is certainly the simplest and most convenient, one finds one really does not discover much about this process. Water consists of hydrogen and oxygen; the two must be conceived as mixed together in the water somehow, but no inward, experiential concept can be formed of this. It is commonly explained in a very external way. Hydrogen consists of atoms (or molecules if you will) and so does oxygen. These intermingle, collide, and cling to one another, and so forth. This means that, although the inner experience was forgotten, the same situation did not arise as in physics, where one could measure things (and increasingly physics became a matter of measuring, counting and weighing). Instead, one could only hypothesize what was happening, the inner process. In a certain respect, this has remained the case in chemistry to this day, because what is pictured as the inner nature of chemical processes is basically only something read into them by thought.

What was then known concerning the inner fluids, that is, how these four fluids—yellow gall, black gall, blood and phlegm—influence and mix with one another, really amounts to an inner human chemistry, though it is of course considered childish today. No other form of chemistry existed in those days. The external phenomena that today belong to the field of chemistry were then evaluated according to these inward experiences. We can therefore speak of an inner chemistry based on experiences of the fluid human being permeated by the ether body. Chemistry was tied to man in former ages. Later it emerged, as did mathematics and physics, and became external chemistry [*see table above*]. Try to imagine how human beings of ancient times experienced physics and chemistry — as a part of themselves, not as something that is mere description of external nature and its processes. Physics and chemistry were really experienced — that is the main point to remember.

## From ritual to experiment

It was not out of some form of childishness, but out of his way of experiencing knowledge that the human being of old came to perform ritualistic ceremonies and to regard them as something real. For he knew that what he created in his ritual was something inward put into outer form, something rooted in a cognition from which he was not estranged, but which connected him to reality. What he impressed into his ritual was what the world had first impressed into him. When he had reached this state of knowledge, he said to himself: Just as the physical breath from the surrounding cosmos lives within me, now the spiritual essence of the world lives in my transformed consciousness. And when I in turn make an outer structure, when I build into the objects and rituals what first formed itself in me out of the spiritual cosmos, I am performing an act that has a direct connection with the spiritual content of the cosmos.

Thus for the human being of an ancient culture, the outward cultic objects stood before him symbolically in such a way that through them he felt again the original connection with the spiritual entities he had first experienced through ordinary knowledge. He knew that in the elements of the ritual something is concentrated in an outer visible form. This something does not exhaust itself in the outward expression I see before me, for soul-spiritual powers that live in the cosmos are alive in the ritual while it takes place.

What I am relating to you is what went on in the souls of those human beings who as a result of their inner experiences gave form to rituals. One reaches a psychological understanding of such rituals when one is willing to accept the idea of inspired cognition.[9] These things simply cannot be explained in the usual external way. One must enter deeply into man's being and must consider how the various functions of the entire human race developed in sequence—how, for instance, particular rituals developed in a certain epoch. The religious ceremonies of today are actually remnants of something that took form in ancient times and then afterwards became fixed. This is why it is increasingly difficult for people today to understand the reason for religious ritual, for they feel it is no longer a justifiable way of relating to the outer world.

Furthermore, we can see another aspect of how the soul works in the course of mankind's development. Deep knowledge, as I have described, underlies the creation of a ritual or the carrying out of a ritual. But humanity has developed further and another factor has entered in, which still lives more or less in the unconscious. What shows itself most clearly when we reach imaginative cognition is that the nervous system is formed out of our soul-spiritual powers. This too has developed in the course of human history. Particularly since the middle of the fifteenth century, humanity in all its various groups has developed in such a way that this instinctive incorporation of soul-spiritual powers into the nervous system has become stronger than it was formerly. We simply have a stronger intellect today. This is obvious when one studies Plato and Aristotle. Our intellect is organized differently. In my *Riddles of Philosophy*[10] I have demonstrated this through the history of philosophy itself. Our intellectual functioning is different. We simply overwork that element of the soul which has grown stronger in the course of human development. And this element which has grown stronger has also become more independent. The increasing independence of our intellect from the nervous system simply has not reached the attention of the philosophers, or of mankind in general. Because the human being has grown stronger inside, so to say, because he has penetrated his nervous system with a stronger organizing power from the soul-spiritual realm, he feels the need to make use of this intensified intellectual activity in the outer world. In ancient times, knowledge attained inwardly was used in the creation and the exercise of rituals; there was a striving to carry over what had been originally experienced inwardly as knowledge into what was performed outwardly. In the same way today, the longing arises to satisfy our stronger, more independent intellect in the outer world. The intellect desires a counterpart that corresponds to the ritual.

What is the result of such a wish? Please accept the paradox, for psychologically it is so: where inner experience is expelled, as it were, where the intellect alone wishes to arrange a procedure so that it can live in the outer object just as cosmic life was once intended to live in the 'object' of the ritual, what results from this is the scientific experiment. Experiment is the way the modern human being satisfies his now stronger intellect. Thereby he lives at the opposite pole from the time when people fulfilled their relation to the cosmos through the cultic object and ritual ceremony. These are the two opposite poles. In an ancient culture of instinctive clairvoyance, the impulse was to give outer form and presence to inner cosmic experience, in what could be called ritualistic exercise. Our intensified modern intellect, on the other hand, is such that it wishes to externalize itself in controlled movements that are devoid of all inwardness, in which nothing subjective lives — and yet the experiment is controlled precisely through the subjective attainments of our intellect. It may seem strange to you that the same underlying impulse gives rise on the one hand to the ritual, and on the other to the experiment, but one can understand these polarities if one considers the human being as a whole.

# 4. Are there Limits to what Science can Know?

*Any field of knowledge functions within certain boundaries, and this is true of scientific knowledge. However, according to Rudolf Steiner, the limits are self-imposed, and can be transcended if an appropriate methodology is found. In his book* The Philosophy of Freedom, *he asserts that knowledge in general is the result of perceptions ('percepts') uniting with concepts. Percepts can always be increased – even beyond the boundaries of the physical world into the spiritual domain. And appropriate concepts can follow. In principle, therefore, there are no limits to knowledge. This is in sharp contrast to thinkers such as Du Bois-Reymond (discussed in this chapter) and Kant. The type of thinking that erects barriers to knowledge can be overcome through Goethe's methodology.*

If we wish to comprehend nature, we must permeate it with concepts and ideas. Why must we do that? Because only thereby does consciousness awake, because only thereby do we become conscious human beings. Just as each morning upon opening our eyes we achieve consciousness in our interaction with the external world, so essentially did consciousness awake within the evolution of humanity. Consciousness, as it is now, was first kindled through the interaction of the senses and thinking with the outer world. We can observe the historical development of consciousness in the interaction of man's senses with outer nature. In this process consciousness gradually was kindled out of the dull, sleepy community and culture of primordial times. Yet one need only consider with an open mind this fact of consciousness, this interaction between the senses and nature, in order to observe something extraordinary transpiring within the human being. We must look into our soul to see what is there, either by remaining, on awakening, within that dull and dreamy consciousness for a while or by looking back into the almost dreamlike consciousness of primordial times. If we look within our soul at what lies submerged beneath the surface consciousness that arises through interaction between our senses and the outer world, we find a world of representations, faint, diluted to dream-pictures with hazy contours, each image fading into the other. Unprejudiced observation ascertains this. The faintness of such images, the haziness of their contours, the fading of one representation into another — none of this can cease unless we awake to a full interaction with external nature. In order to come to this awakening, which is tantamount to becoming fully human, our senses must awake every morning to contact with nature. In the same way it was also necessary for humanity as a whole to awake out of a dull, dreamlike vision of primordial worlds within the soul to achieve the present clear representations and images of the natural world.

By achieving this clarity of representation and the sharply delineated concepts that we need in order to remain awake, we become aware of our environment with a waking soul. We need to awaken in this way in order to remain human in the fullest sense of the word. But we cannot simply conjure this wakefulness out of ourselves. We achieve it only when our senses come into contact with nature; only then do we achieve clear, sharply delineated concepts. We thereby develop something that man must develop for his own sake—otherwise consciousness would not awake. It is thus not an abstract 'need for explanations', not what Du Bois-Reymond[11] and other people like him call 'the need to know the causes of things', that drives us to seek explanations, but the need to become human in the fullest sense through observing nature. We thus may not say that we can outgrow the need to explain, like any other child's play, for that would mean that we would not want to become human in the fullest sense of the word—that is to say, would not want to awake in the way we must awake.

Something else happens in this process, however. In coming to such concepts as we achieve through contemplating nature, we impoverish our inner conceptual life at the same time. Our concepts become clear, but their compass becomes diminished, and if we consider exactly what it is we have achieved by means of these concepts, we see that it is an external, mathematical-mechanical lucidity. Within that lucidity, however, we find nothing that allows us to comprehend life. We have, as it were, stepped out into the light but lost the very ground beneath our feet. We find no concepts that allow us to typify life, or even consciousness, in any way. In exchange for the clarity we must seek for the sake of our humanity, we have lost the content of what we strove for. And then we contemplate nature around us with our concepts. We formulate such complex ideas as the theory of evolution and the like. We strive for clarity. Out of this clarity we formulate a world view, but within this world view it is impossible to find ourselves, to find the human being. With our concepts we have moved out to the periphery, the surface of things, where we come into contact with nature. We have achieved clarity, but along the way we have lost man. We move through nature, apply a mathematical-mechanical explanation, apply the theory of evolution, formulate all kinds of biological laws. We explain nature; we formulate a view of nature — within which man cannot be found. The abundance of content that we once had has been lost, and we are confronted with a concept that can be formed only with the clearest but at the same time most desiccated and lifeless thinking — the concept of matter. And in the face of the concept of matter we are forced to confess that we have achieved clarity, have struggled through to an awakening of full consciousness, but have lost the essence of man in our thinking, in our explanations and comprehension.

And now we turn to look within. We turn away from matter to consider the inner realm of consciousness. We see how within this inner realm of consciousness images and representations pass in review, feelings come and go, impulses of will flash through us. We observe all this and notice that when we attempt to bring the inner realm into the same kind of focus that we achieved with regard to the external world, it is impossible. We seem to swim in an element that we cannot bring into sharp contours, that continually fades in and out of focus. The clarity for which we strive in response to outer nature simply cannot be achieved within us. In the most recent attempts to understand this inner realm, in Anglo-American association psychology, we see how, following the example of Hume, Mill, James, and others, the attempt was made to impose the clarity attained in observation of external nature upon inner sensations and feelings. One attempts to impose clarity upon sensation, and this is impossible. It is as though one wanted to apply the laws of flight to swimming. One does not come to terms at all with the element within which one has to move. The psychology of association never achieves sharpness of contour or clarity about the phenomenon of consciousness. And even if one makes a sober attempt, as Herbart has done, to apply mathematical computation to human mental activity, to the human soul, one finds it possible, but these computations hover in the air. There is no place to gain a proper grasp or foothold, because mathematical formulae simply cannot comprehend what is actually occurring in the soul. While one loses the human being in coming to clarity about the external world, one finds him to be sure—it goes without saying—one finds the human being when one delves into human consciousness; but there is no hope of achieving clarity in this realm, for one swims about, borne hither and thither in an insubstantial realm. One finds the human being, but one cannot find a valid image of what he really is.

It was this that Du Bois-Reymond felt very clearly but was able to express only much less clearly — only as a kind of vague feeling about scientific research in general — when in August 1872 he expressed his 'ignorabimus' or 'we do not know'. What this 'ignorabimus' implies in essence is that on the one hand, in the historical evolution of humanity, we have arrived at clarity about nature and have constructed the concept of matter. In this view of nature we have lost the human being, that is, ourselves. On the other hand we delve into consciousness. To this realm we want to apply what has been most important in arriving at a contemporary explanation of nature. Consciousness rejects this lucidity. This mathematical clarity is entirely out of place. To be sure, we find the human being in a sense, but our consciousness is not yet strong enough, not yet intense enough to comprehend man fully.

Again, one is tempted to answer with an 'ignorabimus', but that cannot be, for we need something more than this in order to meet the social demands of the modern world. The limits that Du Bois-Reymond had come up against when he expressed his 'ignorabimus' on 14 August 1872 lies not within the human condition as such but only within its present stage of historical human evolution. How are we to transcend this statement of failing to know? That is the burning question. It must be answered, not to meet a human 'need to know' but to meet man's universal need to become fully human.

## Where does thinking go wrong?

We have arrived at an indication of what happens when we begin to correlate our consciousness with an external natural world of the senses. Our consciousness awakens to clear concepts but loses itself. It loses itself to the extent that one can only posit empty concepts such as 'matter', concepts that then become enigmatic. Only by thus losing ourselves, however, can we achieve the clear conceptual thinking we need to become fully human. In a certain sense we must first lose ourselves in order to find ourselves again out of ourselves. Yet now the time has come when we should learn something from these phenomena. And what can one learn from these phenomena? One can learn that, although clarity of conceptual thinking and perspicuity of mental representation can be won by man in his interaction with the sense world, this clarity of conceptual thinking becomes useless the moment we strive scientifically for something more than mere empiricism. It becomes useless the moment we try to proceed towards the kind of phenomenalism that Goethe the scientist cultivated, the moment we want something more than natural science, namely, Goetheanism.

What does this mean? In establishing a correlation

between our inner life and the external physical world of the senses we can use the concepts we form in interaction with nature in such a way that we try not to remain within the natural phenomena but to think on beyond them. We do this if we go further than stating, for instance, that within the spectrum the colour yellow appears next to the colour green, and on the other side come the blues — if, instead of simply interrelating phenomena with the help of our concepts we seek to pierce the veil of the senses, as it were, and use our concepts to construct something else above or beyond this veil. We do this if we say: Out of the clear concepts I have achieved I shall construct atoms, molecules — all the movements of matter that are supposed to exist behind natural phenomena. Something extraordinary happens as a result. What happens is that when I, as a human being, confront the world of nature [*see Fig. 1*], I use my concepts not only to create for myself a conceptual order within the realm of the senses but also to break through the boundary of sense and construct behind it atoms and the like. I cannot bring my lucid thinking to a

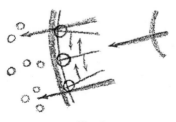

*Fig. 1*

halt within the realm of the senses. I take my lesson from inert matter, which continues to roll on or reverberate even when the propulsive force has ceased. My knowledge reaches the world of the senses, while I remain inert.

I have a certain inertia, and I roll with my concepts on beyond the realm of the senses to fabricate a world there — the existence of which I can begin to doubt when I notice that my thinking has only been borne along by inertia.

It is interesting to note that a great proportion of the philosophy that does not remain within phenomena is actually nothing other than just such an inert rolling-on beyond what really exists in the world. One simply cannot come to a halt. One wants to think further and further on, and construct atoms and molecules — under certain circumstances other things as well that philosophers have assembled there. No wonder, then, that this web one has woven in a world created by the inertia of thinking must eventually unravel itself again.

Goethe rebelled against this law of inertia. He did not want to roll onward with his thinking in this way, but rather to come to a strict halt at this boundary [*see illustration: heavy line*] and to apply concepts *within* the realm of the senses. He thus would say to himself: Within the spectrum appear to me yellow, blue, red, indigo, violet. If, however, I permeate these appearances of colour with my world of concepts while remaining within the phenomena, then the phenomena order themselves of their own accord, and the phenomenon of the spectrum teaches me that when the darker colours or anything dark is placed behind the lighter colours or anything light, there appear the colours which lie towards the blue end of the spectrum. And conversely, if I place light behind dark, there appear the colours which lie towards the red end of the spectrum.

What was it that Goethe was actually seeking to do? Goethe wanted to find simple phenomena within the complex but above all such phenomena as allowed him to remain within this limit [*see illustration*], so that he did not 'roll on' into a realm that one reaches only through a certain mental inertia. Goethe wanted to maintain a strict phenomenalism, keep faith with the phenomena themselves.

## Goethe's use of thinking: reading nature like a book

No doubt you have had some experience of what could be called phenomenalism in the sense of a Goethean world view. In arranging experiments and observations, Goethe used the intellect differently from the way it is used in recent phases of modern thought. He used the intellect as we use it in reading. When we read, we form a whole out of individual letters. For instance, when we have a row of letters and succeed in inwardly grasping the whole, then we have solved a certain riddle posed by this row of individual letters. We would not think of saying: Here is a 'b', an 'r', an 'e', an 'a' and a 'd' — and now I will focus on the 'b'. This isolated 'b' tells me nothing in particular, so I have to penetrate further to what really lies behind the 'b'. Then one could say that behind this 'b' is concealed some mysterious 'beyond', a 'beyond' that makes an impression on me and explains the 'b' to me. Of course, I do not do this; I simply take a look at the succession of letters in front of me and out of them form a whole: I read 'bread'. Goethe proceeds in the same way in regard to the individual phenomena of the outer world. For instance, he does not take some light phenomenon and begins to philosophize about it, wondering what states of vibration lie behind this phenomenon in some sort of 'beyond'. He does not use his intellect to speculate what might be hiding behind the phenomenon; rather, he uses his intellect as we do when we 'think' the letters together into a word. Similarly he uses the intellect solely as a medium in which phenomena are grouped — grouped in such a way that in their relation to one another they let themselves be 'read'. So we can see that Goethe employs the intellect as what I would call a cosmic 'reading tool' to explore the external, physical-mineral, phenomenological world.

He never speaks of a Kantian 'thing in itself' that must be sought behind the phenomena, something Kant supposed existed there. And so Goethe comes to a true understanding of phenomena — of what might be called the 'letters' in the mineral-physical world. He starts with the archetypal or 'Ur'-phenomenon, and then proceeds to more complex phenomena which he seeks either in observation or in the experiments that he contrives. He 'reads' what is spread out in space and time, not looking behind the phenomena, but observing them in such a way that they cast light on one another, expressing themselves as a whole.

His other use of the intellect is to arrange experimental situations that can be 'read'—to arrange experimental situations and then see what is expressed by them. When we adopt such a way of observing phenomena and make it more and more our own, proceeding even further than Goethe, we acquire a certain feeling of kinship with these phenomena. We experience our close inner connection with the phenomena. We enter into them with intensity, in contrast to the way the intellect is used to pierce through the phenomena and seek for all kinds of things behind or underlying them—things which, fundamentally, are only fabricated theories. Naturally, what I have just said only concerns this theoretical activity. We need to educate ourselves in phenomenology, to grow together with the phenomena of the world around us.

# 5. Understanding Organisms: Goethe's Method

*It is generally assumed by most scientists that there is no fundamental distinction between the non-living (inorganic nature) and the living (organic nature), the difference being one of complexity only. Indeed, plants and animals are frequently regarded as very sophisticated machines. This view is not without merit, for there are indeed many complex mechanisms within organisms which yield to machine-like explanations, such as the mechanism for walking which can be mimicked by a robot. Nevertheless, the whole organism, self-regulating, self-repairing, self-reproducing, belongs to a totally different order of phenomena than non-living systems. Goethe insisted that a different methodology must be applied to the study of animals and plants, compared with the inorganic realm; it is out of keeping with reality to apply methods from one realm in another realm which follows different laws. A similar error is current in popular views of the computer, when the attempt is made to understand psychological processes in terms of programming. The mind cannot be understood as a computer, any more than the computer could be understood as a clockwork mechanism. Each realm has its own laws.*

*The extract which follows is from Rudolf Steiner's introduction to Goethe's writings on organic morphology. Its lucidity and the fullness of explanations have ensured that Goethe's approach has not been lost. Meticulous observation of phenomena is also totally in accord with modern scientific method, even if Goethe then allows the phenomena to speak for themselves rather than adding speculative theories.*

Goethe's morphological works are so important because they establish the theoretical foundations and methodology for studying organic nature—a scientific accomplishment of the highest order. To properly appreciate this fact, we must, above all, consider the tremendous difference between inorganic and organic phenomena. The collision of two billiard balls, for example, is an inorganic phenomenon. If one ball is at rest and another strikes it from a certain direction with a certain velocity, the one at rest will begin to move in a certain direction with a certain velocity. We can comprehend such phenomena only by transforming into concepts what is given directly to the senses. We are able to do this to the extent that nothing sense-perceptible remains that we have not penetrated conceptually. We watch one ball approach and strike the other and see the other move on. We have comprehended this phenomenon when—from the mass, direction and velocity of the first ball, and from the mass of the second— we are able to predict the velocity and direction of the second, in other words, when we see that under given conditions this phenomenon complies with necessity. This, however, means only that what presents itself to our senses must appear as a necessary consequence of what we postulate as idea. If this is the case, then we can say that concept and phenomenon coincide. There is nothing in the concept that is not also in the phenomenon, and nothing in the phenomenon that is not also in the concept.

Let us consider more closely the conditions that necessarily lead to an occurrence in inorganic nature. Here we encounter the significant fact that conditions determining sense-perceptible occurrences in inorganic nature also belong to the sensory world. In our example, mass, velocity and direction come into consideration—conditions that in fact belong to the sensory realm. No other conditions determine the phenomenon; only factors directly perceptible to the senses determine one another. A conceptual understanding of such occurrences thus simply deduces tangible reality from tangible reality. Spatial and temporal factors—mass, weight or sense-perceptible forces such as light or warmth—all invoke phenomena belonging to the same category. A body is heated and thereby increases in volume; both the cause and the effect, the heating as well as the expansion, belong to the sensory world.

Thus, we do not have to go beyond the sensory world to grasp such occurrences. We simply deduce one phenomenon from another within that world. When we wish to explain such a phenomenon and understand it conceptually therefore, we need to include only those elements that can be perceived by the senses; everything we wish to understand can be perceived. Therein lies the coincidence of percept (appearance) and concept. Nothing in such events remains obscure to us, because we know the conditions that led to them. Thus we have developed the essential nature of the inorganic world and, in doing so, have shown the extent to which we can explain it through itself, without going beyond it. There has never been any question about this since human beings first began to think about the nature of such things. They did not, of course, always follow the above train of reasoning that leads to the coincidence of concept and percept, but they never hesitated to explain phenomena through their own essential nature as decribed.

This was not true, however, with regard to phenomena in the organic realm until Goethe. The sense-perceptible aspects of the organism — its form, size, colour, temperature, and so on — do not seem to be determined by the same kind of factors. One cannot say, for example, that the size, form, position, and so on, of a plant's roots determine the sense-perceptible characteristics of the leaf or blossom. A body in which this was the case would not be an organism, but a machine. Rather, it must be granted that the sense-perceptible characteristics of a living being do not appear as a result of other sense-perceptible conditions as in inorganic nature.[12] Indeed, all sensory qualities arise in an organism as the result of something no longer perceptible the senses. They appear as the result of a higher unity hovering over sense-perceptible processes. It is not the form of the root that determines that of the stem, and not the form of the stem that determines that of the leaves, and so on. All of these forms are determined instead by something that exists beyond them and whose form is inaccessible to the senses. The perceptible elements exist for one another, but not as a result of one another. They are not mutually determined by one another but by something else. Here what we perceive with our senses cannot be reduced to other sensory factors; we must include in our concept of events elements that do not belong to the world of the senses; we must go beyond the sensory world. What we perceive is no longer enough; to comprehend the phenomena, we must conceptually grasp the unifying principle. Consequently, however, a distance arises between percept and concept, and they no longer seem to coincide; the concept hovers over what is observed and it becomes difficult to see how they are connected.

Whereas concept and sensory reality were one in inorganic nature, they seem to diverge here and belong, in fact, to two different worlds. What is perceived and presents itself directly to the senses no longer seems to bear its own explanation, or essence, within itself. The object does not seem to be self-explanatory because its concept is taken from something other than itself. Because the object is present to the senses but does not seem to be governed by the laws of the sensory world, it is as though we were confronted by an insoluble contradiction in nature. It is as though there were a gulf between inorganic phenomena, which are self-explanatory, and organic beings, where natural laws are intruded upon, and legitimate laws seem suddenly to be broken. This chasm was in fact generally taken for granted in science until Goethe, who succeeded in solving this mystery. Until then, it was assumed that only inorganic nature could be explained through itself; the human capacity for knowledge was believed to end at the level of organic nature.

We can comprehend the magnitude of Goethe's achievement when we consider the fact that Kant, the great reformer of recent philosophy, not only shared that old erroneous concept fully, but even sought for a scientific reason why the human mind would never be able to explain organic entities. He did, in fact, acknowledge the possibility of an intuitive intellect (*intellectus archetypus*) capable of grasping the connection between concept and sense reality both in organic beings and in the inorganic realm. But he denied humankind the possibility of such an intellect. According to Kant, the human intellect can conceive of the unity, or concept, of a thing only as arising from the interaction of its parts, as an analytical generalization arrived at through abstract reasoning, but not in such a way that each part appears as the result of a definite, concrete (or synthetic) unity — that is, as the result of an intuitive concept. Consequently, he considered it impossible for the human intellect to explain organic nature, the activity of which must be viewed as emanating from the whole into the parts.

This is the essential thing: an occurrence in inorganic nature, that is, something that takes place only in the sensory world, is caused and determined by a process that likewise occurs only in the sensory world. Imagine that the causal process consists of elements m, d and v (mass, direction and velocity of a moving billiard ball), and the resulting process of elements m', d' and v'. Whenever m, d and v are given, m', d' and v' will be determined by them. If I wish to understand this occurrence, I must formulate the entire event, consisting of cause and effect, with one concept that includes both. But this kind of concept cannot exist within the occurrence itself and determine it. It encompasses both processes in a common expression; but it does not cause or determine. Only the objects of the sensory world determine one another. The elements m, d and v are also perceptible to the outer senses. The concept, in this case, simply serves to summarize the external event; it expresses something that is not real as an idea or concept, but is real to the senses. This 'something' it expresses is an object of sensory perception. Knowledge of inorganic nature is based on the possibility of comprehending the external world through the senses and expressing its interactions through concepts. Kant regarded the possibility of knowing things in this way as the only kind of knowledge accessible to the human being. He called this kind of thinking discursive. What we wish to know is an external perception; the concept, or combining unity, is merely a *means*.

If, however, we wish to comprehend organic nature, then, according to Kant, we cannot apprehend the ideal, conceptual aspect as something that borrows its meaning by expressing or indicating something else; rather, we would have to apprehend the ideal element as such. It would have to contain its own meaning, stemming from itself, not from the spatial-temporal world of the senses. The unity that our mind merely abstracts in the case of the inorganic would have to build upon itself, forming itself out of itself; it would have to be fashioned according to its own being, not according to the influences of other objects. According to Kant, the comprehension of such a self-forming and self-manifesting entity is denied to the human being.

What is needed to attain such comprehension? We need a kind of thinking that can give thought a substance not derived from outer sensory perception, a thinking that comprehends not only what is perceived externally by the senses, but also apprehends pure ideas apart from the sensory world. A concept that is not abstracted from the sensory world but whose content develops out of itself and only out of itself can be called an intuitive concept, and the comprehension of such a concept may be called 'intuitive knowledge'. What follows from this is clear: a living organism can be comprehended only through an intuitive concept. Goethe actually demonstrated that it *is* in fact possible for us to know in this way.

The inorganic world is governed by the interaction of the individual elements comprising an event, by the reciprocal way they determine one another. This is not true of the organic world, where one member of an organism does not determine another, but where the whole (or idea) determines each particular out of itself in accordance with its own being. In referring to this self-determining entity, we can follow Goethe's terminology and call it an entelechy. An entelechy, therefore, is a force that calls itself into being. The resulting manifestations have sensory existence, but they are determined by this entelechy principle. This creates an apparent contradiction: the organism is self-determining and creates its characteristics out of itself according to a postulated principle, and yet it has sense-perceptible reality. A living organism thus attains sense-perceptible reality in a way that is completely different from other objects of the sensory world. Consequently, it seems to arise unnaturally.[13]

It is also entirely understandable that, externally, an organism is exposed to the effects of the sensory world, as is any other body. A tile falling from the roof can strike a living creature as well as an inorganic object. An organism is related to the outer world through its intake of nourishment and so on; it is affected by the physical circumstances of the outer world. This can, of course, occur only to the degree that the organism is an object of the spatial-temporal world of the senses. This object of the outer world — the entelechy principle that has manifested outwardly — is the organism's external appearance. Consequently, it seems neither to fully accord with itself nor to adhere strictly to its own nature. But since the organism is subject not only to its own formative laws but to conditions of the outer world as well, since it is not only as it should be according to its own self-determining entelechy principle but also what it has become through the influence of other factors upon which it depends, it never seems to fully accord with itself nor to adhere wholly to its own nature.

This is where human reason enters and, within the realm of ideas, develops an organism that corresponds only to its own principle, setting aside influences from the outer world. Every incidental influence that has nothing to do with the organic as such thus falls away completely. This idea, which corresponds purely to the organic aspect of the organism, is the archetypal organism — it is Goethe's 'type'.

Thus the eminent validity of the idea of the type becomes apparent. It is not merely an intellectual concept but the truly organic aspect of every organism, without which it would not be an organism. Because it manifests in every organism it is more real than any actual, particular organism. It also expresses the essence of an organism more fully and purely than any individual organism in particular. The way we arrive at the idea of the type is fundamentally different from the way we arrive at the concept of an inorganic process, which is abstracted from external reality and is not active within it. The idea of the organism, on the other hand, works actively within the organism as its entelechy — it is the essence of the entelechy itself in a form apprehended by our reason. The idea is not a summary of experience; it produces experience. Goethe expresses this as follows: 'A concept is the sum of experience, an idea is its result; to understand a concept requires intellect, to comprehend an idea requires reason.' This explains the kind of reality that must be attributed to

Goethe's archetypal organism (the archetypal plant or animal). This Goethean method is clearly the only way we can apprehend the essential nature of the organic world.[14]

The organism presents itself to us in nature in two main forms—as plant and as animal, each in a different way. The plant distinguishes itself from the animal through its lack of any real inner life. In an animal, this inner life manifests as sensation, intentional movement, and so on. The plant has no such soul principle. It does not go beyond the development of its external form. As the entelechy principle unfolds its formative activity out of one point, so to speak, it manifests in the plant through the fact that each organ is shaped according to a common formative principle. The entelechy appears here as the force that forms the individual organs. The organs are all shaped according to one formative type, and they appear as modifications of one fundamental organ; they are repetitions of that organ at various developmental levels. What makes the plant a plant, a particular formative force, is active in every organ in the same way. In this sense each organ is identical with all the others and also with the plant as a whole. Goethe expressed it this way:

> I have come to realize that the organ of the plant we ordinarily call the leaf conceals the true Proteus, who can conceal and reveal himself in all formations. Backward and forward, the plant is only leaf, linked so inseparably to the future seed that one should not think one without the other.[15]

Thus the plant appears to be made up of many individual plants, so to speak, like a complex individual made up of less complex individuals. The development of the plant progresses from stage to stage and forms its organs; each organ is identical to every other in its formative principle, though different in appearance. The plant's inner unity spreads itself into outer breadth; it expresses and loses itself in diverse forms, so that it does not (as occurs in the animal) attain its own concrete existence and a certain independence, which then, as a centre of life, encounters the multiplicity of its organs and uses them as mediators with the outer world.

We must now ask what causes the external differentiation of plant organs, which are otherwise identical in terms of their inner principle? How can formative laws, guided by a single formative principle, produce a leaf in one instance and a sepal in another? Since plants exist entirely in the external realm, this differentiation must be based on outer, spatial factors. Goethe considered alternating expansion and contraction to be such factors. When the plant's entelechy principle enters outer existence, working from a point outwards, it manifests as a spatial entity. The formative forces are active in space and create organs of a specific spatial form. Now these forces either concentrate and draw inwards to a single point during a stage of contraction or they spread by unfolding, striving as it were to distance themselves from one another during a stage of expansion. Over the course of the plant's life, three expansions alternate with three contractions. This alternating expansion and contraction causes the essentially identical formative forces of the plant to differentiate.

Initially, the whole potential of the plant—contracted to a single point—is dormant in the seed [*a of Fig. 2*]. It then emerges, unfolds, and expands in the formation of the leaves (*c*). The formative forces increasingly repel one another; consequently, the lower leaves appear rudimentary and compact (*cc′*); further up, they become more ribbed and indented. All that was crowded together begins to diverge [*leaves d and e*]. All that was previously separated by successive intervals (*zz′*) appears—as the formation of the calyx (*f*)—drawn into a single point on the stalk (*w*). This is the second contraction. In the corolla of the blossom a new unfolding, or expansion, takes place.

Compared with the sepals (*f*) of the calyx, the petals (*g*)

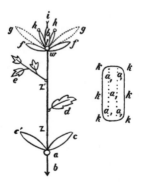

Fig. 2

are finer and more delicate, which could be caused only by diminishing contraction towards one point, that is, by a greater extension of the formative forces. The next contraction takes place within the reproductive organs, the stamens (*h*) and pistil (*i*). After that a new expansion begins with the formation of the fruit (*k*). In the seed (*a1*) that emerges from the fruit, the whole being of the plant is again condensed to a single point.[16]

The entire plant represents the unfolding, or realization, of the potential of the bud or seed, which requires only the right outer influences to fully develop as a plant form. The only difference between a bud and a seed is that the seed requires the earth as its ground, whereas the bud generally represents the formation of a plant upon a plant. The seed represents an individual plant of a higher nature, or a whole cycle of plant formations, so to speak. With each new bud, the plant begins a new phase of its life; it regenerates itself and concentrates its forces to renew them. Consequently, the formation of a bud interrupts the vegetative process. The life of the plant can withdraw into a bud when the conditions for manifest life are lacking, and sprout again when the right conditions reappear. This is why vegetative growth is interrupted during the winter. Goethe said of this:

> It is very interesting to observe how vegetation continues its vigorous growth when it is not interrupted by severe cold; here [in Italy] there are no buds, and one begins to understand what a bud actually is.[17]

Thus, what in our climate lies hidden in the bud is there openly displayed. Indeed, true plant life lies hidden within it; it lacks only the conditions for its unfolding.

Goethe's concept of alternating expansion and contraction has encountered especially strong opposition. All of those attacks, however, stem from a misunderstanding — the belief that these concepts cannot be valid unless a physical cause can be found for them, and unless it is possible to demonstrate the means whereby the inner laws of the plant cause this expansion and contraction. This, however, places the cart before the horse. Nothing should be postulated as the cause of expansion and contraction. On the contrary, everything else follows from them; they themselves cause the progressive metamorphosis, stage by stage.

Such misunderstandings arise whenever we fail to apprehend a concept in its own intuitive form; we then insist that it must be the result of an external occurrence. We can think of expansion and contraction only as caused and not as causing. Goethe did not regard expansion and contraction as resulting from inorganic processes in the plant; rather, he saw them as the way in which the plant's inner entelechy principle shapes itself. Thus, he could not view them as a summation of sense-perceptible processes from which they could be deduced, but was obliged to see them as resulting from the inner unitary principle itself.

A plant's life is maintained by its metabolism. There is a fundamental difference between the metabolism of those organs closer to the roots which take in nourishment from the earth, and those receiving nourishment that has passed through other organs. The organs closer to the earth seem to depend directly on their inorganic environment; the other organs, on the other hand, depend on the parts of the organism that precede them. Each succeeding organ thus receives nourishment especially prepared for it as it were by the preceding organs. Nature progresses from seed to fruit in successive stages, so that what follows manifests as the result of what came before. Goethe referred to this step-by-step progression as 'progress on a spiritual ladder'. Nothing more than what we have indicated lies in his words:

> ... because an upper node arises from the preceding one and receives the juices mediated by it, a node higher on the stem must receive its juices in a finer and more filtered state, and it must benefit from the previous development of the leaves, refine its form, and direct a still finer sap to its leaves and buds.

We begin to understand all of these things when we see them in the light of Goethe's ideas. The ideas presented here are the elements that lie in the nature of the archetypal plant in a way that corresponds only to the archetype itself, and not as they manifest in any particular plant, where they do not appear in their original form but adapted to outer conditions.

In animal life, of course, something else intervenes. The life of the animal does not lose itself in its outer features, but separates itself, severs itself from its corporeality and uses its bodily manifestation only as a tool. It no longer manifests merely as the capacity to shape the organism from within; rather it expresses itself within the organism as something beside the organism, acting as its governing force. The animal manifests as a self-contained world, or microcosm, in a much higher sense than does the plant. It has a centre that is served by all its organs.

> Thus every mouth is adept at grasping the fodder
> Fit to the body's need, whether weak and toothless its
>    jaw,
> Or mighty with menacing teeth; in every case
> An organ perfectly suited provides for all other organs.
> Also each foot and leg, be it a long or a short one,
> Moves to serve with great skill the impulse and need of
>    the creature.[18]

Each organ of the plant contains the whole plant, but the life principle exists nowhere as a definite centre; the identity of the organs lies in the fact that they are all formed according to the same laws. In the animal, each organ appears to proceed from this definite centre; the centre shapes all organs according to its own nature. Thus the shape of the animal provides the basis for its outer existence, but it is determined from within. The animal's way of life is therefore directed by those same inner formative principles. The inner life of the animal, on the other hand, is free and unrestricted within itself; within certain limits, it can adapt to outer influences, but it is determined by the inner nature of the type and not by outer, mechanistic influences. In other words, adaptation cannot go so far as to cause the organism to appear as a mere product of the outer world. Its formation is confined within certain limits:

> No god can extend these limits, since they are honoured by nature; perfection was never achieved except through such limitation.[19]

If every animal conformed only to the principles of the archetypal animal, all animals would be the same. But the animal organism is differentiated into several organ systems, each of which has the capacity for a certain degree of development. This forms the basis for differentiated evolution. Ideally they are all equally important, but one system may nevertheless predominate and draw the organism's whole store of formative forces to itself and away from the other organs. Such an animal will appear especially developed in the direction of that organ system, whereas another animal will develop in another way. This gives rise to the possibility for the archetypal organism to differentiate as the various genera and species when it enters the phenomenal realm.

The actual (factual) causes of this differentiation have not been mentioned yet. This is where external factors come into play — adaptation, through which the organism shapes itself in terms of its outer environment, and the struggle for existence, which allows only those creatures to survive that adapt most successfully to prevailing conditions. Adaptation and the struggle for existence, however, would have no effect at all upon the organism if its formative principle could not assume diverse forms while maintaining its inner unity. We should not imagine that this principle is influenced by external formative forces in the same way that one inorganic entity affects another. The outer conditions are indeed responsible for the fact that the type assumes a particular form; the form itself, however, is not derived from those outer conditions but from the inner principle. When explaining the form we must always consider outer conditions, but we should not think that the form itself results from them. Goethe would have rejected the idea that environmental influences lead to organic forms simply through causality, just as he rejected the teleological principle that explains the form of an organ in terms of an external purpose.

In the organ systems of an animal that are related more to its outer structure (the skeleton, for example) we see a law observed in plants appearing again, for example in the formation of the bones of the skull. Goethe's talent for seeing the inner lawfulness within purely external forms is especially evident here.

This distinction between plants and animals based on Goethe's views might seem irrelevant, given the fact that recent science has found good reason to doubt that there is any fixed boundary between plants and animals. But even Goethe was aware that it is impossible to establish any such boundary. This did not prevent him from clearly defining the plant and the animal, in a way that relates to his world view as a whole. Goethe assumed that absolutely nothing is constant or fixed in the phenomenal realm where everything continually fluctuates and moves. The essence of a thing, however, which we grasp in concepts, cannot be derived from fluctuating forms but from certain intermediate stages in which it can be observed. The way Goethe views the world naturally establishes certain definitions, which, nevertheless, are not rigidly maintained in the face of certain transitional forms. In fact, this is exactly where Goethe sees the flexibility of natural life.

With the ideas described here, Goethe laid the theoretical foundation for organic science. He discovered the essential nature of the organism. It is easy to miss this fact if we assume that the type (the principle that forms itself out of itself, the entelechy) can be explained in terms of something else. But such an assumption is unjustified, because the type, when apprehended intuitively, is self-explanatory. Anyone who has understood this 'forming itself according to itself' of the entelechy principle will see this as the solution to the mystery of life. Any other solution is impossible, because this is the essence of the matter. If Darwinism is compelled to postulate a primal organism, then we can say that Goethe discovered the essential nature of that primal organism.[20]

It was Goethe who broke with merely classifying genera and species and initiated a regeneration of organic science in keeping with the true nature of the organism. Whereas systematizers before Goethe required as many different concepts, or ideas, as there are externally differentiated species (between which they could find no transition), Goethe declared all organisms to be the same in idea, differing only in appearance. And he explained why this is so. Thus a philosophical basis was created for a scientific system of organisms; all that remained was to elaborate it. In what sense all existing organisms are only revelations of an idea and how they reveal it in particular cases would have to be demonstrated.

The type, which is self-generating, has the capacity to assume endlessly diverse forms as it manifests. Such forms are the objects of our sensory perception, the genera and species of organisms actually living in space and time. To the degree that our mind has understood that general idea—the type—it has also comprehended the entire world of organisms in its unity. When we then see the type as it has taken shape in each particular phenomenal form, those forms become intelligible; they appear as stages, as metamorphoses, in which the type realizes itself. Essentially, the task of the new systematic science based on Goethe's insights is to point out these various stages.

In both the animal and the plant kingdoms, an ascending evolutionary sequence prevails; organisms are differentiated by their degree of development. How is this possible? We can characterize the ideal form, or type, of organisms by the fact that it is made up of spatial and temporal elements. Consequently, it appeared to Goethe as a 'sensory-supersensory' form. It contains spatial-temporal forms that can be perceived as idea (intuitively). When it manifests in the phenomenal realm, the actual

sense-perceptible form (now no longer intuitively perceived) may or may not correspond completely to the
ideal form; the type may or may not attain its full development. Certain organisms are lower because their
phenomenal forms do not correspond fully to the organic
type. The more the outer appearance and the organic type
coincide in a particular being, the more perfect it is. This is
the objective reason for an ascending evolutionary
sequence. It is the task of any systematic presentation to
investigate this relationship within the form of each
organism. When establishing the type—the primal or
archetypal organisms—however, this cannot be taken into
consideration; one must find a form that represents the
most perfected expression of the type. Goethe's archetypal
plant provides such a form.

# 6. The Quest for Archetypal Phenomena

*Is it possible to imagine a science which does not classify phenomena, does not look for causes, nor laws of nature? In Goethe's approach each of these factors assumes a particular dynamic form which requires more active participation by the human spirit, in contrast to habitual thought patterns. Phenomena are sorted and combined in such a way that they mutually illuminate each other, without the need for 'laws' or theoretical explanations in terms of inferred entities such as atoms. The difference between living and non-living then becomes more evident. Following this method, and phrasing it here in terms of elementary physics, Rudolf Steiner is able to point to two categories of force: the normal 'centric' forces such as magnetism, electricity, gravity, which are active in that part of nature which is relatively lifeless; and 'peripheral' or cosmic forces, which cannot be calculated, and which are active in living nature. Elsewhere Steiner describes peripheral forces as etheric formative forces. In doing so he is not inventing causes — that would be un-Goethean! Rather, in faithfulness to Goethe's method, he is merely including what to him is perceptible.*

The scientist today seeks to approach nature from three vantage points. In the first place he is at pains to observe nature in such a way that from her many creatures and phenomena he may form concepts of species, kind and genus. He subdivides and classifies the beings and phenomena of nature. You need only recall how in external, sensory experience so many single wolves, single hyenas, single phenomena of warmth, single phenomena of electricity present themselves to the human being, who then attempts to gather up the single phenomena into kinds and species. So then he speaks of the species 'wolf' or 'hyena', likewise he classifies phenomena into species, grouping and classifying what is given, to begin with, in many single instances and experiences. Now we may say that this first important activity is already taken more or less unconsciously for granted. Scientists in our time do not reflect that they should really examine how these 'universals', these general ideas, are related to the single data.

The second thing done by people of today in scientific research is that they try by experiment, or by conceptual elaboration of the results of experiment, to arrive at what they call the 'causes' of phenomena. Speaking of causes, our scientists will have in mind forces or substances, or even more universal factors. They speak for instance of the force of electricity, the force of magnetism, the force of heat or warmth, and so on. They speak of an unknown 'ether'[21] or the like, as underlying the phenomena of light and electricity. From the results of experiments they try to arrive at the properties of this ether. Now you are well aware how very controversial is all that can be said about the 'ether' of physics. There is one thing however to which we may draw attention even at this stage. In trying, as they put it, to go back to the causes of phenomena, scientists are always trying to find their way from what is known into some unknown realm. They scarcely ever ask if it is really justified to proceed in this way from the known to the unknown. They scarcely trouble, for example, to consider whether it is justified to say that when we perceive a phenomenon of light or colour what we subjectively describe as the quality of colour is the effect on us, upon our soul, our nervous apparatus, of an objective process that is taking place in the universal ether — say a wave-movement in the ether. They do not pause to think whether it is justified to distinguish (which is what they really do) between the 'subjective' event and the 'objective', the latter being the supposed wave-movement in the ether, or else the interaction of it with processes in quantifiable matter.

Shaken though it now is to some extent, this kind of scientific outlook was predominant in the nineteenth century; and we still find it everywhere, in the whole way phenomena are spoken of. It still undoubtedly prevails in scientific literature to this day.

Now there is also a third way in which the scientist tries to grasp nature's workings. He takes the phenomena to begin with — for instance a simple phenomenon such as the fact that a stone, let go, will fall to earth, or if suspended by a string will pull vertically down towards the earth. Phenomena like this the scientist sums up and so arrives at what he calls a 'law of nature'. The following statement, for example, would be regarded as a simple law of nature: 'Every celestial body attracts to itself the bodies that are upon it.' We call the force of attraction gravity or gravitation and then express how it works in certain 'laws'. Another classical example are the three statements known as 'Kepler's Laws'.

It is in these three ways that 'scientific research' tries to get close to nature. Now I will emphasize at the very outset that the Goethean outlook on nature strives for the very opposite in all three respects. In the first place, when Goethe began to study natural phenomena, the classification into species and genera — whether of the creatures or of the facts and phenomena of nature — at once became problematical for him. He did not like to see the many specific entities and facts of nature reduced to all these rigid concepts of species, family and genus; what he desired was to observe the gradual transition of one phenomenon into another, or of one form of manifestation of an entity into another. He felt concerned not with the subdivision and classification into genera, but with the metamorphosis both of phenomena and of the different creatures. The quest for so-called 'causes' in nature, which science has gone on pursuing ever since Goethe's time, was not according to his way of thinking either. In this respect it is especially important for us to realize the fundamental difference between natural science and research as pursued today, and on the other hand the Goethean approach to nature.

The science of our time makes experiments; having thus

studied the phenomena, it then tries to form ideas about so-called causes that are supposed to underlie them. Behind the subjective phenomenon of light or colour, for example, scientists seek an objective wave-movement in the ether. Goethe did not apply scientific thinking in this way. In his researches into nature he does not try to proceed from the so-called 'known' to the so-called 'unknown', but always tries to stay within the sphere of what is known; nor in the first place is he concerned to enquire whether the latter is merely subjective or objective. Goethe does not entertain such concepts as the 'subjective' phenomena of colour and the 'objective' wave-movements in outer space. What he beholds spread out in space and going on in time is for him one, a single undivided whole. He does not question whether it is subjective or objective. His use of scientific thinking and scientific method is not to draw conclusions from the known and apply them to the unknown; instead he applies all thinking and all available methods to combine the phenomena themselves, until, in the last resort, he reaches the phenomena which he calls archetypal, the Ur-phenomena. These archetypal phenomena — regardless of 'subjective or objective' — bring to expression what Goethe feels is fundamental to a true outlook on nature and the world. Goethe therefore remains embedded in the sequence of actual phenomena; he only sifts and simplifies them, and then calls 'Ur-phenomenon' the simplified and clarified phenomenon, ideally transparent and comprehensive.

Thus Goethe looks upon the whole of scientific method — so to call it — purely and simply as a means of grouping and combining phenomena. Staying with the actual phenomena, he wants to group them in such a way that they themselves express their secrets. Nowhere does he seek to extrapolate from the so-called 'known' to some 'unknown'. Hence for Goethe, in the last resort, there are not what may properly be called 'laws of nature' either. He is not seeking such laws. What he puts down as the quintessence of his researches are simple facts — the fact, for instance, of how light will interact with matter that is in its path. Goethe puts into words how light and matter interact. That is no 'law'; it is a pure and simple fact. And upon facts like this he seeks to base his contemplation, his whole outlook on nature. What he desires, fundamentally, is a rational description of nature. Only for him there is a difference between the mere crude description of a phenomenon as it may first present itself, where it is still complex and opaque, and the description which emerges when one has sifted it, so that the simple essentials alone emerge. This then — the Ur-phenomenon — is what Goethe takes to be fundamental, in place of the unknown entities or the conceptually defined 'laws' of our customary science.

One fact may throw considerable light on what is seeking to enter our science via Goetheanism, and on what now obtains in science. It is remarkable: few people have ever had so clear an understanding of the relation of natural phenomena to mathematical thinking as Goethe had. Goethe himself not having been much of a math-ematician, this is disputed no doubt. Some people think he had no clear idea of the relation of natural phenomena to those mathematical formulations which have grown ever more beloved in science, so much so that in our time they are felt to be the one and only firm foundation. Increas-ingly in modern times, the mathematical way of studying natural phenomena — I do not actually say the mathemat-ical study of nature, it would not be right to put it in these words, but the study of natural phenomena in terms of mathematical formulae — has grown to be the determining factor in the way we think even of nature itself.

We really do need to find clarity about these things. You see, my friends, when approaching nature in the cus-tomary way, we have three things to begin with — things that are really practised by man before he gets to grips with nature. The first is common or garden arithmetic. In studying nature nowadays we do a lot of arithmetical counting and calculating. Arithmetic — we must be clear on this — is something man understands on its own ground, in and through itself. When we are counting, it makes no difference what we count. Learning arithmetic, we receive something which, to begin with, has no refer-ence to the outer world. We may count peas as well as electrons. The way we recognize that our methods of counting and calculating are correct is altogether different from the way we contemplate and form conclusions about the outer processes to which our arithmetic is then applied.

The second of the three to which I have referred is again a thing we do before we come to outer nature. I mean geometry — all that is known by means of pure geometry. What a cube or an octahedron is, and the relations of their angles — all these are things which we determine without looking into outer nature. We spin and weave them out of ourselves. We may make outer drawings of them, but this is only to serve mental convenience, not to say inertia. Whatever we may illustrate by external drawings, we might equally well imagine purely in the mind. Indeed it is very good for us to imagine more of these things purely in the mind, using the crutches of outer illustration rather less. In other words, what we have to say about geometrical form is derived from a realm which, to begin with, is quite distant from external nature. We know what has to be said about a cube without first having had to deduce it from a cube of rock salt. Yet we must discover it in the latter. So we ourselves do something quite separate from nature and then apply it to the latter.

And then there is the third thing which we do, again before reaching external nature. I am referring to what we do in 'phoronomy' so-called, or kinematics, i.e. the science of movement. Now it is very important for you to be clear on this point — to realize that kinematics too is, fundamentally speaking, still remote from what we call the 'real' phenomena of nature. Imagine an object to be moving from point a to point b [*Fig. 3a*]. I am not looking at any moving object; I just imagine it. Then I can always imagine this movement from a to b, indicated by an arrow in the

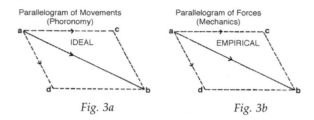

Fig. 3a          Fig. 3b

figure, to consist of two distinct movements. Think of it like this. Point a is ultimately to arrive at b, but we suppose it does not go there at once. It sets out in this other direction and reaches c. If it then subsequently moves from c to b, it does eventually get to b. So I can imagine the movement from a to b proceeding not along the line a–b but along the line, or the two lines, a–c–b. The movement ab then consists of the movements ac and cb, i.e. of two distinct movements. You need not observe any process in outer nature; you can simply think it—picture it to yourself in thought—how the movement from a to b is composed of the two other movements. That is to say, in place of the one movement the two other movements might be carried out with the same ultimate effect. And when in thinking I picture this, the thought—the mental picture— is spun out of myself. I need have made no outer drawing. I could simply have instructed you in thought to form the mental picture; you would have to find it valid. Yet if in outer nature there is really something like point a—perhaps a little ball, a grain of shot—which in one instance moves from a to b and in another from a to c and then from c to b, what I have pictured to myself in thought will really happen. This is how it is in kinematics, in the science of movement; I think the movements to myself, yet what I think proves applicable to the phenomena of nature and must indeed hold good there.

So we really can say that in arithmetic, in geometry and in phoronomy or kinematics we have the three preliminary steps that precede our actual study of nature. Spun as they are purely out of ourselves, the concepts that we gain in all these three are nonetheless valid for what actually takes place in nature.

And now I beg you to remember the so-called parallelogram of forces [*Fig. 3b*]. This time, point a signifies a material thing—some little grain of material substance. I exert a force to draw it from a to b. Notice the difference between the way I am now speaking and the way I spoke before. Before, I spoke of movement as such; now I am saying that a force draws the little ball from a to b. Suppose the measure of this force, pulling from a to b, to be five grams; you can denote it by a corresponding length in this direction. With a force of five grams I pull the little ball from a to b. I might also do it differently—I might first pull with a certain force from a to c. Pulling from a to c (with a force denoted by this length) I need a different force than when I pulled directly from a to b. Then I might add a second pull, in the direction of the line from c to b, and with a force denoted by the length of this line. Having pulled in the first instance from a towards b with a force of five grams, I should have to calculate from this figure how

big the pull a–c and also how big the pull c–d would have
to be. Then if I pulled simultaneously with forces represented by the lines a–c and a–d of the parallelogram, I
should be pulling the object along in such a way that it
eventually got to b; thus I can calculate how strongly I
must pull towards c and d respectively. Yet I cannot calculate this in the same way as I did the displacements in
our previous example. What I found previously (in relation to the movement pure and simple), it was possible
simply to calculate purely in thought. Not so when a real
pull, a real force is exerted. Here I must somehow measure
the force; I must approach nature itself; I must proceed
from thought to the world of facts. If once you realize this
difference between the parallelogram of movements and
that of forces, you have a clear and sharp formulation of
the essential difference between all those things that can
be determined within the realm of thought and those that
lie beyond the range of thoughts and mental pictures. You
can reach movements but not forces with your mental
activity. Forces you have to measure in the outer world.
The fact that when two pulls come into play — the one
from a to c, the other from a to d — whether the object is
actually pulled from a to b according to the parallelogram
of forces can only be ascertained by external experiment.
There is no proof by dint of thought, as for the parallelogram of movements. It must be measured and ascertained
externally. So in conclusion we may say that while we
derive the parallelogram of movements by pure reasoning, the parallelogram of forces must be derived empiri-

cally, by dint of outer experience. Distinguishing the parallelogram of movements and that of forces, you have the difference—clear and sharp—between phoronomy and mechanics, or kinematics and mechanics. Mechanics has to do with forces, not mere movements; it is already a natural science. Mechanics is concerned with the way forces work in space and time. Arithmetic, geometry and kinematics are not yet natural sciences in the proper sense. To reach the first of the natural sciences, which is mechanics, we have to go beyond the life of ideas and mental pictures.

Even at this stage our contemporaries fail to think clearly enough. I will explain by an example how great is the leap from kinematics to mechanics. Kinematic phenomena can still take place entirely within our own thinking; mechanical phenomena on the other hand must first be tried and tested by us in the outer world. Our scientists, however, do not envisage the distinction clearly. They always tend to confuse what can still be seen in purely mathematical ways, and what involves realities of the outer world. What, in effect, must be there, before we can speak of a parallelogram of forces? So long as we are only speaking of the parallelogram of movements, no actual body need be there; we need only imagine one in our thinking. For the parallelogram of forces on the other hand there must be a mass—a mass that possesses weight among other things. This you must not forget. There must be a mass at point a, to begin with. Now we may well feel driven to enquire what mass actually is. What is it really?

And we shall have to admit that we already get stuck at this question! The moment we take leave of things which we can settle purely in the world of thought so that they then hold good in outer nature, we get into difficult and uncertain regions. You are of course aware how scientists proceed. Equipped with arithmetic, geometry and kinematics, to which they also add a little dose of mechanics, they try to work out a mechanics of molecules and atoms; for they imagine what is called matter to be classifiable in this way. In terms of this molecular mechanics they then try to conceive the phenomena of nature, which, in the form in which they first present themselves, they regard as our own subjective experience.

We take hold of a warm object, for example. The scientist will tell us: What you call its heat or warmth is the effect on your own nerves. Objectively, there is a movement of molecules and atoms. These you can study, following the laws of mechanics. So then they study the laws of mechanics, of atoms and molecules. Indeed, for a long time they imagined that by so doing they would at last contrive to explain all the phenomena of nature. Today, of course, this hope is rather shaken. But even if we do reach the atom with our thinking, even then we shall have to ask—and seek the answer by experiment—What are the forces in the atom? How does the mass reveal itself in its effects? How does it work? And if you put this question, you must ask again: How will you recognize it? You can only recognize the mass by its effects.

The customary method in answering this question is to

recognize the smallest unit-bearer of mechanical force by its effect. If such a particle brings another minute particle — say a minute particle of matter weighing one gram — into movement, there must be some force proceeding from the matter in the one, which brings the other into movement. If the given mass then brings the other mass, weighing one gram, into movement in such a way that the latter goes a centimetre a second faster in each successive second, the former mass will have exerted a certain force. This force we are accustomed to regard as a kind of universal unit. If we are then able to say of some force that it is so many times greater than the force needed to make a gram go a centimetre a second quicker every second, we know the ratio between the force in question and the chosen universal unit. If we express it as a weight, it weighs 0.001019 grams. Indeed, to express what this kind of force involves, we must have recourse to the balance — the weighing-machine. The unit force is equivalent to the downward thrust that comes into play when 0.001019 grams are being weighed. So then I have to express myself in terms of something very outwardly real if I want to approach what is called 'mass' in this universe. Howsoever I may think it out, I can only express the concept 'mass' by introducing what I get to know in quite external ways, that is, weight. In the last resort, it is through weight that I express the mass. And even if I then go on to atomize it, I still express it as a weight.

I have reminded you of all this in order to describe clearly the point at which we pass from what can still be

determined a priori into the realm of actual nature. We
need to be very clear on this point. The truths of arithmetic, geometry and kinematics we can undoubtedly
determine separately from external nature. But we must
also be clear to what extent these truths are applicable to
what meets us, in effect, from quite another angle — and, to
begin with, in mechanics. Not until we get to mechanics
do we have the content of what we call 'natural
phenomena'.

All this was clear to Goethe. Only where we pass from
kinematics to mechanics can we begin to speak at all of
natural phenomena. Aware as he was of this, he knew that
this is the only possible relation of mathematics to natural
science, however much mathematics is idolized in the
scientific domain.

To stress this point I will cite one further example. Even
as we may think of the unit element, for the effects of force
in nature, as a minute atom-like body that would be able
to impart an acceleration of a centimetre per second per
second to a gram weight,[22] so too with every manifestation of force we shall be able to say that the force proceeds
from one direction and works towards another. Thus we
may well grow accustomed — for all the workings of
nature — always to look for the points from which the
forces proceed. Precisely this has grown habitual, or even
dominant, in science. Indeed in many instances we really
find it so. There are whole fields of phenomena which we
can thus refer to the points from which the forces dominating the phenomena proceed. We therefore call such

forces 'centric forces', inasmuch as they always issue from central points. It is indeed right to think of centric forces wherever we can find a certain number of single points from which quite definite forces, dominating a given field of phenomena, proceed. Now need the forces always come into play? It may well be that the point-centre in question only bears in it the possibility, the potentiality as it were, for such a play of forces to arise, whereas the forces do not actually come into play until the requisite conditions are fulfilled in the surrounding sphere. We shall find instances of this. It is as though forces were concentrated at the points in question, forces however that are not yet in action. Only when we bring about the necessary conditions will they call forth actual phenomena in their surroundings. Yet we must recognize that forces are concentrated in such points or space, and potentially able to work on their environment.

This in effect is what we always look for when describing the world in terms of physics. All physical research amounts to this: we pursue the centric forces to their centres; we try to find the points from which effects can issue. For this kind of effect in nature we are obliged to assume that there are centres, charged as it were with possibilities of action in certain directions. And we have sundry means of measuring these possibilities of action; we can express in specified measures how strongly such a point or centre has the potentiality of working. Speaking in general terms, we call the measure of a force thus centred and concentrated a 'potential' or 'potential force'.

In studying these effects of nature we then have to trace the potentials of the centric forces, as we may formulate it; we look for centres which we then investigate as sources of potential forces.

Such, in effect, is the line taken by that school of science which is at pains to express everything in mechanical terms. It looks for centric forces and their potentials. In this respect we need to take one essential step — out into actual nature — in order to grow fully conscious of the fact that you cannot possibly understand any phenomenon in which life plays a part if you restrict yourself to this method, looking only for the potentials of centric forces. Say you were studying the play of forces in an animal or vegetable embryo or germ-cell — with this method you would never find your way. No doubt it seems an ultimate ideal to the science of today, to understand even organic phenomena in terms of potentials, of centric forces of some kind. It will be the dawn of a new world-conception in this realm when it is recognized that the thing cannot be done in this way. Phenomena in which life is active can never be understood in terms of centric forces. Why not? Diagrammatically, let us here imagine that we are setting out to study transient, living phenomena of nature in terms of physics. We look for centres, to study the potential effects that may proceed from such centres. Suppose we find the effect. If I now calculate the potentials, say for the three points a, b and c, I find that a will work thus and thus on A, B and C; or c on A', B' and C' and so on. I should thus get a notion of how the integral effects will, in a certain sphere, be subject to the potentials of such and such centric forces. Yet in this way I could never explain any process that involves life. The forces that are essential to a living thing have no potential; they are not centric forces. If at a given point d you tried to trace the physical effects due to the influences of a, b and c, you would indeed be referring to the effects of centric forces, and you could do so. But if you want to study the effects of life you can never do this. For these effects, there are no centres such as a or b or c. Here your thinking will only take the right direction when you look at things in the following way. Say that at d there is something alive. I look for the forces to which the life is subject. I shall not find them in a, nor in b, nor in c, nor when I go still farther outwards. I only find them when as it were I go to the very ends of the world — and, what is more, to the entire circumference at once. Taking my start from d, I should have to go to the outermost ends of the universe and imagine forces working inwards from the spherical circumference from all sides, forces whose interplay unite in d. It is the very opposite of the centric forces with their potentials. How to calculate a potential for what works inwards from all sides, from the infinities of space? In the attempt I should have to dismember the forces — one total force would have to be divided into ever-smaller portions. Then I should get nearer and nearer the edge of the world; the force would be completely sundered, and so would all my calculation. Here in effect it is not centric forces, but cosmic, universal forces that are at work. And here calculation ceases.

Once more you have a leap—this time from what is not alive in nature to what is. In the investigation of nature we shall only find our way if we know what the leap is from kinematics to mechanics, and again what the leap is from external, inorganic nature into those realms that are no longer accessible to calculation—where every attempted calculation breaks down and every potential evaporates. This second leap will take us from external inorganic nature into living nature, and we must realize that calculation ceases where we want to understand what is alive.

Now in this explanation I have been neatly dividing all that refers to potentials and centric forces and on the other hand all that leads out into cosmic forces. Yet in the nature that surrounds us they are not divided like this. You may ask where we can find an object where only centric forces work with their potentials, and on the other hand where is the realm where cosmic forces work, which do not let you calculate potentials. An answer can indeed be given, one that reveals the very great importance of what is involved here. For we may truly say: All that man makes in the shape of machines—all that is pieced together by man from elements supplied by nature—represents the purely centric forces working according to their potentials. What exists in nature outside us on the other hand—even in inorganic nature—can never be attributed exclusively to centric forces. In nature there is no such thing; it never works completely in that way. Save in the things made artificially by man, the workings of centric and cosmic forces are always flowing together in their effects. In the whole realm of so-called nature nothing is truly 'un-living'. The one exception is what man makes artificially — man-made machines and mechanical devices.

The truth of this was profoundly clear to Goethe, as a natural instinct, and his whole outlook on nature was founded on this. Here we have the quintessence of the contrast between Goethe and the modern scientist as represented by Newton. The scientists of modern times have only looked in one direction, always observing external nature in such a way as to attribute all things to centric forces — as it were to expunge everything in nature which cannot be defined in terms of centric forces and their potentials. For Goethe such an outlook was not sufficient. What was viewed as 'nature' in this way seemed to him a void abstraction. There is reality for him only where centric forces and peripheral or cosmic forces are both involved — where there is interplay between the two. On this polarity, in the last resort his *Theory of Colour* is also founded, of which we shall speak in more detail.

# 7. Light, Darkness and Colour

*Any study of Goethean science, and of Steiner's extension of it to include the super-sensory realm, would be incomplete without the Colour Theory. Goethe's view is based entirely on natural phenomena and easily reproducible experimental observations. No theoretical explanation is sought (such as waves or corpuscles); rather, the phenomena are arranged and analysed so as to identify the simplest (or primary) components, which provide the key to all the others. In his 'Light Course' Steiner demonstrated many of the experiments to the teachers of the first Waldorf School in Stuttgart. Reading his descriptions, a scientist may well experience some difficulty in keeping an open mind towards descriptions of phenomena without interposing more familiar theories. The contrast with a more conventional approach could not be greater. For example, Goethe and Steiner regard 'darkness' as a reality in its own right, not merely a lack of light. However, when it comes to practical applications and predictions, both approaches can hold their own. Yet the Goethean approach has the advantage of leaving the door open to a deepening of knowledge into the spiritual realm.*

Now the fact is that all we encounter as colour really confronts us in two opposite and polar qualities, no less than magnetism does, to take another example, in positive

and negative magnetism. There is no less of a polar quality
in the realm of colour. At the one pole is all that which we
describe as yellow and the kindred colours—orange and
reddish. At the other pole is what we may describe as blue
and the kindred colours—indigo and violet and even
certain shades of green. Why do I emphasize that we
encounter the world of colour as a polar quality? Because
in fact the polarity of colour is among the most significant
phenomena of all nature and should be studied accord-
ingly. To proceed at once to what Goethe calls the Ur-
phenomenon in the sense I was explaining yesterday, this
is indeed the Ur-phenomenon of colour. We shall discover
it to begin with by looking for colour in and about the light
as such. This is to be our first experiment, arranged as well
as we are able. I will explain first what it is. The experiment will be as follows:

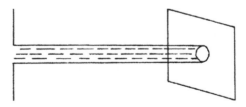

*Fig. 4*

Through a narrow slit—or a small circular opening, we
may assume to begin with—in an otherwise opaque wall,
we let in light [*Fig. 4*]. We let the light pour in through the
slit. Opposite the wall through which the light is pouring

in, we put a screen. By virtue of the light that is pouring in, we see an illuminated circular surface on the screen. The experiment is best done by cutting a hole in the shutters, letting the sunlight pour in from outside. We can then put up a screen and catch the resulting image. We cannot do it in this way right now, so we are using the lantern to project it. When I remove the shutter, you see a luminous circle on the wall. This, to begin with, is the picture that arises through the fact that a cylinder of light, passing along here, is caught on the opposite wall. We now put a prism into the path of this cylinder of light [*Fig. 5*].

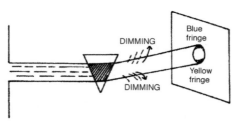

*Fig. 5*

The light can then no longer simply penetrate to the opposite wall and there produce a luminous circle; it is compelled to deviate from its path. How have we brought this about? The prism is made of two planes of glass, set at an angle to form a wedge. This hollow prism is then filled with water. We let the cylinder of light, produced by the projecting apparatus, pass through the water-prism. If you now look at the wall, you see that the patch of light is

no longer down there, where it was before. It is displaced—it appears elsewhere. Moreover you see a peculiar phenomenon—at its upper edge you see a bluish-
greenish light. You see the patch with a bluish edge
therefore. Below, you see the edge is reddish-yellow.

This then is what we have to begin with—the
'phenomenon'. Let us first stay with the phenomenon,
simply describing the fact as it confronts us. In going
through the prism, the light is somehow deflected from its
path. It now forms a circle higher up, but if we measured it
we should find it is not an exact circle. It is drawn out a
little above and below, and edged with blue above and
yellowish below. If therefore we cause such a cylinder of
light to pass through the prismatically formed body of
water—neglecting, as we can in this case, whatever
modifications may be due to the plates of glass—colour
phenomena arise at the edges.

Now I will do the experiment again with a far narrower
cylinder of light. You see a far smaller patch of light on the
screen. Deflecting it again with the help of the prism, once
more you see the patch of light displaced—moved
upwards. This time however the circle of light is completely filled with colours. The displaced patch of light
now appears violet, blue, green, yellow and red. Indeed, if
we made a more thorough study of it, we should find all
the colours of the rainbow in their proper order. We take
the fact, purely and simply as we find it. And please, all
those of you who at school learned those neatly finished
diagrams with rays of light, normals[23] and so on, please

forget them now. Stay with the simple phenomenon, the pure and simple fact. We see colours arising in and about the light and we can ask ourselves what is it due to. Look please once more; I will again insert the larger aperture. There again is the cylinder of light passing through space, impinging on the screen and there forming its picture of light [*Fig. 4*]. Again we put the prism in the way. Again the picture of light is displaced and the phenomena of colour appear at the edges [*Fig. 5*].

Now please observe the following. We will remain purely within the given facts. Kindly observe. If you could look at it more exactly you would see the luminous cylinder of water where the light is going through the prism. This is a matter of simple fact: the cylinder of light goes through the prism of water and there is thus an interpenetration of the light with the water. Pay careful attention please, once more. Because the cylinder of light passes through the water, the light and water inter-penetrate, and this is evidently not without effect for the surroundings. On the contrary, we must aver (and once again, we add nothing to the facts in saying this), that the cylinder of light somehow has power to make its way through the water-prism to the other side, yet in the process it is deflected by the prism. Were it not for the prism, it would go straight on, but it is now thrown upwards and deflected. Here then is something that deflects our cylinder of light. To denote what is deflecting our cylinder of light by an arrow in the diagram, I shall have to put the arrow thus [*see diagram above*]. So we can say, adhering once again to the facts and not indulging in speculations: By means of such a prism the cylinder of light is deflected upwards, and we can indicate the direction in which it is deflected.

And now, to add to all this, think of the following, which once again is a simple statement of fact. If you let light pass through a dim and milky glass or through any cloudy fluid — through dim, cloudy, turbid matter in effect — the light is of course dimmed. When you see light through clear, unclouded water, you see it in full brightness; if the water is cloudy, you see it weakened. By dim and cloudy media the light is weakened; you will see this in countless instances. We have to state this, to begin with, simply as a fact. Now in some respects, however little, every material medium has a clouding effect. So does this prism here. It always dims the light to some extent. That is to say, with respect to the light that is there within the prism, we are dealing with a light that is somehow dimmed. Here to begin with [*pointing to Fig. 4*] we have the light as it shines forth; here on the other hand we have the light that has made its way through the material medium. In here however, inside the prism, matter and light work together; a dimming of the light arises here. That the dimming of the light has a real effect, you can tell from the simple fact that when you look into light through a dim or cloudy medium you see something more. The dimming has an effect — this is perceptible. What comes about through the dimming of light? We have to do not only with the cone of light that is here bent and deflected, but also with this new factor—the dimming of the light, brought about by matter. We can imagine therefore that into this space beyond the prism not only the light is shining, but there shines in, rays in to this light the quality of dimness that is in the prism. How does it ray in? Naturally it spreads out and extends after the light has gone through the prism. What has been dimmed and darkened rays into what is light and bright. You need only think of it carefully and you will admit that the dimness too is shining up into this region here. If what is light is deflected upwards, then what is dim is deflected upwards too. That is to say, the dimming is deflected upwards in the same direction as the light is. The light that is deflected upwards has a dimming effect, so to speak, sent after it. Up there the light cannot spread out unimpaired, but into it the darkening, the dimming effect is also sent. Here then we are dealing with the interaction of two things: the brightly shining light, itself deflected, and then the sending into it of the darkening effect that is poured into this shining light. The dimming and darkening effect is here deflected in the same direction as the light is. And now you see the outcome. Here in this upper region the bright light is infused and irradiated with dimness, and by this means the dark or bluish colours are produced.

How is it then when you look further down? The dimming and darkening shines downwards too, naturally. But you see what happens. While there is a part of the outraying light where the dimming effect takes the same direction as the light that surges through—so to speak—

with its prime force and momentum, here on the other hand the dimming effect that has arisen spreads and shines further, so that there is a space in which the cylinder of light as a whole is still diverted upwards. Yet at the same time, into the body of light which is thus diverted upwards, the dimming and darkening effect rays in too. Here is a region where, through the upper parts of the prism, the dimming and darkening goes downwards. Here therefore we have a region where the darkening is deflected in the opposite sense — opposite to the deflection of the light. Up there, the dimming or darkening tends to go into the light; down here, the working of the light is such that the deflection of it works in an opposite direction to the deflection of the dimming, darkening effect. This, then, is the result: above, the dimming effect is deflected in the same sense as the light; thus in a way they work together. The dimming and darkening gets into the light like a parasite and mingles with it. Down here on the contrary, the dimming rays back into the light but is overwhelmed and as it were suppressed by the latter. Here therefore, even in the battle between bright and dim — between the lightening and darkening — light predominates. The consequences of this battle. The consequences of the mutual opposition of the light and dark, and of the dark being irradiated by the light, are in this downward region the red or yellow colours. So therefore we may say: Upwards, the darkening runs into the light and there arise the blue shades of colour; downwards, the light outdoes and overwhelms the darkness, and there arise the yellow shades of colour.

You see, dear friends, simply through the fact that the prism on the one hand deflects the full bright cone of light, and on the other hand also deflects the dimming of it, we have two ways in which the dimming or darkening play into the light—the two kinds of interplay between them. We have an interplay of dark and light, not getting mixed to give a grey but remaining mutually independent in their activity. But at one pole they remain active in such a way that the darkness comes to expression as darkness even within the light, whilst at the other pole the darkening works counter to the light—it remains there and independent, it is true, but the light overwhelms and outdoes it. So the lighter shades arise—all that is yellowish in colour. Thus by observing the plain facts and simply taking what is given, purely from what you see, you have the possibility of understanding why yellowish colours on the one hand and bluish colours on the other make their appearance. At the same time you see that the material prism plays an essential part in the way colours arise. For it is through the prism that, on the one hand, the dimming effect is deflected in the same direction as the cone of light, while on the other hand, because the prism lets its darkness ray there too, the darkness raying on and the light that is deflected cut across each other. For that is how the deflection works down here. Below, the darkness and light interact in a different way than above.

Colours therefore arise where dark and light work together. This is what I wanted to make clear to you today. Now if you want to consider for yourselves how to best understand this, you need only think for instance of how differently your own etheric body is integrated into your muscles and into your eyes. Into a muscle it is integrated in such a way as to blend with the functions of the muscle; not so into the eye. The eye being very isolated, here the etheric body is not integrated into the physical apparatus in the same way, but remains comparatively independent. Consequently, the astral body[24] can come into very intimate union with the portion of the etheric body within the eye. Inside the eye our astral body is more independent, and independent in a different way than in the rest of our physical organization. If we consider the physical organization of a muscle and of the eye we must say that our astral body is integrated into both, but in a very different way. In the muscle it is integrated in such a way that it goes through the same space as the physical bodily part and is by no means independent. In the eye too it is integrated; here however it works independently. The space is filled by both, in both cases, but in the one case the constituents work independently while in the other they do not. It is but half the truth to say that our astral body is there in our physical body. We must ask *in what way* it is there, for in the eye and in the muscle it is present in quite different ways. In the eye it is relatively independent, and yet it is present—no less than in the muscle. You see from this that constituents can interpenetrate each other and still be independent. So too, you can unite light and dark to get grey; then they are interpenetrating like astral body and muscle. Or on the other hand light and dark can so interpenetrate as to retain their mutual independence; then they are inter-penetrating as do the astral body and the physical orga-nization in the eye. In the one instance, grey arises; in the other, colour. When they interpenetrate like the astral body and the muscle, grey arises; whilst when they interpenetrate like the astral body and the eye, colour arises, since they remain relatively independent in spite of being present together in the same space.

## The primary phenomenon of colour

I will begin by placing before you what we may call the 'Ur-phenomenon' or primary phenomenon of the *Theory of Colour*. By and by, you will find it confirmed and reinforced in the phenomena you can observe through the whole range of so-called optics or colour theory. Of course the phenomena get complicated; the simple Ur-phenom-enon is not always easy to recognize at once. But if you take the trouble you will find it everywhere. The simple phenomenon—expressed in Goethe's way, to begin with—is as follows. When I look through darkness at something lighter, the light object will appear modified by the darkness towards the light colours, i.e. in the direction of the red and yellowish tones. If for example I look at anything luminous and, as we should call it, white, at any whitish shining light through a thick enough glass which is in some way dim or cloudy, then what would seem to me more or less white if I were looking at it directly will appear yellowish or yellow-red [*Fig. 6a*].

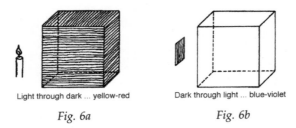

Light through dark ... yellow-red       Dark through light ... blue-violet

*Fig. 6a*            *Fig. 6b*

This is the one pole. Conversely, if you have here a simple black surface and look at it directly, you will see it as black, but if you interpose a trough of water through which you send a stream of light so that the liquid is illumined, you will be looking at the dark through something light. Blue or violet (bluish-red) tones of colour will appear. The other pole is thus revealed. This therefore is the Ur-phenomenon: light seen through dark is yellow; dark seen through light is blue.

This simple phenomenon can be seen on every hand if we once accustom ourselves to think more realistically and not so abstractly as in modern science. Please now recall from this point of view the experiment which we have done. We sent a cylinder of light through a prism and so obtained a real scale of colours, from violet to red; we caught it on a screen. I made a drawing of the phenomenon [*see Fig. 5 and Fig. 7*]. You will remember that if this is

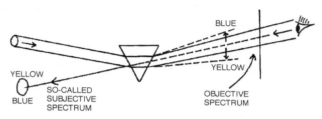

*Fig. 7*

the prism and this the cylinder of light, the light in some way goes through the prism and is diverted upwards. Moreover, as we said before, it is not only diverted. It would be simply diverted if a transparent body with parallel faces were interposed. But we are putting a prism into the path of the light — that is, a body with convergent faces. In passing through the prism, the light is darkened. The moment we send light through the prism we therefore have two things: first the simple light as it streams on its way; and then the dimness interposed in the path of the light. Moreover this dimness, as we said, puts itself into the path of the light in such a way that while the light is mainly diverted upwards the dimming that arises, raying upwards as it does, shines also in the same direction into which the light itself is diverted. That is to say, darkness rays into the diverted light. Darkness is living, as it were, in the diverted light, and by this means the bluish and violet shades are here produced. But the darkness rays downwards too. So while the cylinder of light is diverted upwards, the darkness here rays downwards and works contrary to the diverted light but is no match for it. Here therefore we may say that the original bright light, diverted as it is, overwhelms and outdoes the darkness, and then we get the yellowish or yellow-red colours.

If we now take a sufficiently thin cylinder of light, we can also look in its direction through the prism. Instead of looking from outside onto a screen and seeing the image projected there, we put our eye in the place of this image, and, looking through the prism, we then see, displaced, the aperture through which the cylinder of light is produced [*Fig. 7*]. Once again therefore, adhering strictly to the facts, we have the following phenomenon. Looking along here, I see what would be coming directly towards me if the prism were not there, displaced in a downwards direction by the prism. At the same time I see it coloured.

What then do you see in this case? Watch what you see, state it simply and then connect it with the fundamental fact we have just now been ascertaining. Then, what you actually see will emerge in all detail. Only you must keep to what is really seen. For if you are looking thus into the bright cylinder of light, which once again is now coming towards you, you see something light, namely, the cylinder of light itself, but you are seeing it through dark. (That there is something darkened here is clearly proved by the fact that blue arises in this region.)

Through something darkened — through the blue colour, in effect — you look at something light, namely, at the cylinder of light coming towards you. Through what is dark you look at what is light; here therefore you should be seeing yellow and red — and in fact you do. Likewise the red colour below is proof that here is a region irradiated with light. For as I said just now, the light here overwhelms the dark. Thus as you look in this direction, however bright the cylinder of light itself may be you still see it through an irradiation of light, in relation to which it is dark. Below, therefore, you are looking at dark through light and you will see blue or bluish-red. You need only express the primal phenomenon — it tells you what you actually see. Your eye here encounters what you would be seeing in the other instance. Here for example is the blue and you are looking through it; therefore the light appears reddish. At the bottom edge you have a region that is illumined. However light the cylinder of light may be, you see it through a space that is lit up. Thus you are seeing something darker through an illumined space and so you see it as blue. It is the polarity that matters.

For the phenomenon we studied first — that on the screen — you may use the name 'objective' colours if you wish to speak in learned terms. This other one — the one you see in looking through the prism — will then be called the 'subjective' spectrum. The 'subjective' spectrum appears as an inversion of the 'objective'.

Concerning all these phenomena there has been much intellectual speculation, my dear friends, in modern times. The phenomena have not merely been observed and stated purely as phenomena, as we have been endeavouring to do. There has been a great deal of speculation about them. Indeed, beginning with the famous Newton, science has gone to the uttermost extremes in speculation. Newton, having first seen and been impressed by this colour spectrum, began to speculate about the nature of light. Here is the prism, said Newton; we let the white light in. The colours are already there in the white light; the prism conjures them forth and now they line up in formation. I have then dismembered the white light into its constituents. Newton now imagined that to every colour corresponds a kind of substance, so that seven colours altogether are contained as specific substances in the light. Passing the light through the prism is to Newton like a kind of chemical analysis, whereby the light is separated into seven distinct substances. He even tried to imagine which of the substances emit relatively larger corpuscles — tiny spheres or pellets — and which smaller. According to this conception the sun sends us its light, we let it into the room through a circular opening and it passes through as a cylinder of light. This light however consists of ever so many corpuscles — tiny little bodies. Striking the surface of the prism they are diverted from their original course. Eventually they bombard the screen, where these tiny cannon-balls impinge. The smaller ones fly farther upwards, the larger ones remain farther down. The smallest are the violet, the largest are the red. So then the large are separated from the small.

This idea, that there is a substance or that there are a number of substances flying through space, was seriously shaken before long by other physicists — Huygens, Young and others — until at last physicists said to themselves: The theory of little corpuscular cannon-balls starting from somewhere, projected through a refracting medium or not as the case may be, arriving at the screen and there producing a picture, or again finding their way into the eye and giving rise in us to the phenomenon of red, etc., will not do after all. They were eventually driven to this conclusion by an experiment of Fresnel's, towards which some preliminary work had previously been done by the Jesuit Grimaldi among others.[25]

Fig. 8a                    Fig. 8b

These are the things which I would have you note. A physicist explaining things in Newton's way would naturally argue: If I have something white here—say a luminous strip—and I look at it through the prism, it appears to me in such a way that I get a spectrum: red, orange, yellow, green, blue, dark blue, violet [*Fig. 8a*].

Goethe said: Well, at a pinch, that might do. If nature really is like that—if it has made the light composite—we might well assume that with the help of the prism this light gets analysed into its several parts. Well and good; but now the very same people who say the light consists of these seven colours—so that the seven colours are parts or constituents of the light—these same people allege that darkness is just nothing, is the mere absence of light. Yet if

I leave a strip black in the midst of white, if I have simple white paper with a black strip in the middle and look at this through a prism, then too I find I get a rainbow, only the colours are now in a different order [*Fig. 8b*]: mauve in the middle and on the one side merging into greenish-blue. I get a band of colours in a different order. Following the analytic theory I ought now to say: Then the black too is analysable and I should thus be admitting that darkness is more than the mere absence of light. The darkness too would have to be analysable and would consist of seven colours. This fact — that he saw the black band too in seven colours, only in a different order — was what put Goethe off. And this again shows us how necessary it is simply to take the phenomena as we find them.

# 8. The Rediscovery of the Elements

*Earth, water, air, fire — the well-known 'elements' of ancient Greek thought — were once dismissed as primitive and simplistic in view of the hundred or so 'elements' of modern chemistry. Nowadays most scientists would simply accept them as poetic terms for the solid, liquid and gaseous states, and heat. However, as Rudolf Steiner points out, the terms meant much more, including both earthly and cosmic relationships. Such a cosmic understanding of matter can be recovered in a modern form, as Steiner shows in his 'Warmth Course', given shortly after the 'Light Course'. The central task was to extend Goethe's phenomenological methods in order to understand the true nature of heat. However, as we shall see, heat occupies a central position in nature, and therefore the result of Steiner's research is a comprehensive 'scale of nature' accommodating all the states of matter and several of the forces of nature.*

*The extracts in this chapter contain very simple and well-known considerations of the states of matter, yet culminate in an extremely important methodological breakthrough: that each state contains an image that becomes reality in the next state. This opens the door to research into states beyond the three physical ones, as we shall see in subsequent chapters.*

## What did the Greeks mean?

The real meaning of the ideas and concepts of physical phenomena still prevalent in ancient Greece has been lost. Experimentation began, ideas and concepts were taken up parrot-fashion, without the inner thought process that had accompanied them in ancient Greece. Everything that the Greeks had included in these physical concepts was forgotten. The Greeks had not simply said, 'solid, liquid, gaseous'; what they expressed may be translated into our own language as follows:

Whatever was solid was called earth in ancient Greece
Whatever was fluid was called water in ancient Greece
Whatever was gaseous was called air in ancient Greece

It is totally wrong to think that we can carry our own meaning of the words earth, water and air over into ancient writings in which Greek influence was dominant and assume that the corresponding words have the same meaning there. When we come across the word water in ancient writings, we must translate it with our word *fluidity* and the word earth with our word *solids*. Only in this way can we correctly translate the ancient writings. But a profound significance is implicit in this. The use of the word earth to indicate solids implied especially that this solid condition falls under the laws prevailing on the planet earth. Solids were designated as earth because the Greeks wanted to convey the idea that when a body is solid it is under the influence of earthly laws in every respect. On the other hand, when a body was spoken of as water, then it was not merely under earthly laws but influenced by the entire planetary system. The forces active in fluid bodies, in water, arise not merely from the earth but from the planetary system. The forces of Mercury, Mars, and so on are active in everything that is fluid. But they act in a way arising from the position of the planets, which shows a kind of resultant in the fluidity.[26]

In ancient Greece, therefore, people had the feeling that only solid bodies, designated as earth, were solely under the earthly system of laws, and that when a body melted it was influenced from outside the earth.

And when a gaseous body was called air, the feeling was that such a body was under the unifying influence of the sun. (These things are presented simply in a historical way here.) This gaseous body was lifted out of merely earthly and planetary influence and stood under the unifying influence of the sun. Earthly air beings[27] were regarded in this way; sun forces were essentially active in their configuration, their inner arrangement and substance. Ancient physics had a cosmic character. It was willing to take into account the forces present as facts. For the Moon, Mercury, Mars, etc. are facts. But people lost the origins of this view of things and were at first unable to develop a need for new sources. They therefore could only conceive that since solid bodies in their expansion and in their whole configuration fell under earthly laws, liquid and gaseous bodies must do likewise. You might say that it would never occur to a physicist to deny that the sun warmed the air, etc. Indeed, he does not deny this, but since he proceeds from concepts which delineate the action of the sun according to ideas springing from observations on the earth, he explains the sun in terrestrial terms instead of explaining the terrestrial in solar terms.[28]

The essential point is that consciousness of certain things was completely lost in the period between the fifteenth and seventeenth centuries. Consciousness that our earth is a part of the whole solar system, and that consequently every single thing on the earth has to do with the whole solar system, was lost. The feeling was also lost that the solidity of bodies originated through the earthly element emancipating itself from the cosmic, tearing itself free to attain independent laws, while the lawfulness of the gaseous realm, for example, the air, remained under the unifying influence of the sun as it affected the earth as a whole.[29] This is what has led to the need to explain things terrestrially, things which were formerly given a cosmic explanation. Since man no longer looked to the activity of planetary forces when a solid body changed to a fluid — as when ice becomes fluid, changing to water — forces were no longer sought in the planetary system and had to be placed within the fluid body itself. It was necessary to rationalize and theorize about the way in which atoms and molecules were arranged in such a body. And the inner capacity to bring about the change from solid to liquid, from liquid to gas, had to be ascribed to these unfortunate molecules and atoms. Formerly such a change was considered to act through what was present in space, through cosmic regions beyond the earth.

In this way we must understand the transition of the concepts of physics as shown especially in the crass materialism of the *Accademia del Cimento*,[30] which flowered in the ten-year period between 1657 and 1667. You must picture how this crass materialism arose through the gradual loss of ideas embodying the connection between the earthly and the cosmos beyond the earth. Today we once more face the need to realize this connection. It will not be possible to emerge from materialism unless we cease being pedantic in this field of physics. Narrow-mindedness arises simply because we go from concrete to abstract concepts, for no one loves abstract concepts more than the philistine. He wishes to explain everything with a few formulae, a few abstract ideas. But physics cannot hope to advance if it continues to spin theories as has been the fashion ever since the materialism of the *Accademia del Cimento*. We will progress in our understanding of a field such as heat only if we seek to establish once again the connection between the terrestrial and the cosmic through wider and more comprehensive ideas than modern materialistic physics can furnish us with.

## Solid, liquid and gas between earth and cosmos

We have observed the rise in temperature as we warmed a solid body to melting-point. We showed how the temperature rise disappeared for a time and then reappeared until the body began to boil, to evaporate. When we extended our observations, another fact appeared. We could see that the gas produced spilled over in all directions into its surroundings [*see Fig. 9, left*], seeking to distribute itself in all directions, and could only be made to take on form if its own pressure were opposed by an equal and opposite pressure brought to bear from without. These things have been demonstrated by experiment and will be clarified further by other experiments. The moment the temperature is lowered, to the point where the body can solidify, it can acquire form [*see Fig. 9, centre*]. When we experience temperature simply rising and falling, we experience what corresponds externally to form. We are experiencing the dissolution of form and the re-establishment of it. Gas dissolves form for us, solid establishes form. We experience the transition between these two, and we experience it in an extremely interesting way. Imagine the solid and the gas, and between them the liquid, the fluid body. This liquid need not be enclosed by a vessel surrounding it completely, but only on the bottom

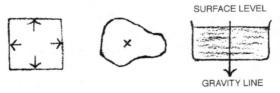

*Fig. 9*

and sides. On the upper side, the liquid forms its own surface perpendicular to the line between itself and the centre of the earth. Thus we can say that we have here a transitional state between the gas and the solid [*see Fig. 9, right*]. In a gas we never have such a surface. In a liquid such as water, there is formed a single surface. In the case of a solid, there is all around the body what in the liquid occurs only on the upper surface.

This is an extremely interesting and significant relationship, for it directs our attention to the fact that the entire surface of a solid body has something corresponding to the upper surface of a liquid; the solid, however, determines the creation of the surface by virtue of its own nature. How does water establish its surface level? It is at right angles to the line joining it to the centre of the earth. The whole earth conditions the creation of this surface. We can therefore say that, in the case of water, each point within it has the same relationship to the entire earth that the points in a solid have to something within the solid. The solid therefore encompasses something that in the case of water resides in the water's relationship to the earth. The gas turns away from the earth, it diffuses. It does not enter into a relationship with the earth at all. It totally eludes this relationship to the earth. Gases have no surface at all.

You will see from this that we are obliged to go back to an old conception. I called your attention to the fact that the ancient Greek physicists called solid bodies 'earth'. They did not do this out of some superficial concept such as has been ascribed to them by people today, but they did it because they were conscious of the fact that the solid, by itself, takes care of something that in the case of water is taken care of by the earth as a whole. The solid takes into itself the role of the earthly. It is entirely justified to put the matter in this way: the earthly resides within a solid. In water it does not entirely reside within, but the whole earth takes up the role of forming a surface on the liquid.

You see, therefore, that when we proceed from solid bodies to water we are obliged to extend our considerations not only to what actually lies before us. In order to get an intelligent idea of the nature of water, we must include the water of the whole earth and to think of this as a unity in relation to the central point of the earth. To observe a 'fragment' of water as a physical entity is as absurd as to consider a cut-off fragment of my little finger as an organism. The fragment would die at once. It only has meaning as an organism if it is considered in its relation to the whole organism. Water does not in itself have the meaning that the solid has in itself. It has meaning only in relation to the whole earth. And so it is with all liquids on the earth.

And again, when we pass on from the fluid to the gaseous, we come to understand that the gaseous removes itself from the earthly domain. It does not form normal surfaces. It partakes of everything that is not terrestrial. In other words, we must not merely look on the earth for what is active in a gas; we must consider what surrounds the earth to help us out, we must go out into space and seek there the forces involved. When we wish to learn the laws of the gaseous state, we become involved in nothing less than astronomical considerations.

Thus you see how these things are related to the whole terrestrial scheme when we examine the phenomena that up to now we have simply gathered together. And when we come to a point such as the melting-point or boiling-point, very significant things arise. For when we consider the melting-point, we pass from the terrestrial condition of a solid body, in which it determines its own form and relations, to something that includes the whole earth. The earth begins to take the solid captive when the solid passes into the fluid state. From its own kingdom, the solid enters the terrestrial kingdom as a whole when we reach the melting-point. It ceases to have individuality. And when we carry the fluid over into the gaseous condition, then we come to the point where its connection with the earth as shown by the formation of a liquid surface is loosened. The instant we go from a liquid to a gas, the body loosens itself from the earthly domain, as it were, and enters the realm of things beyond earth. We must consider the forces active in a gas as having escaped from the earth. When we study these phenomena, therefore, we cannot avoid passing from the ordinary physical-terrestrial into the cosmic. For we are no longer in contact with reality if our attention is not turned to what is actually working in the things themselves. But now we encounter other phenomena. Consider a fact like the one with which you are very familiar and to which I have called your attention, namely, that water behaves very unusually, that ice floats on water, or, stated otherwise, is less dense than water. When it passes into the fluid condition its temperature rises, and it contracts and becomes denser. Only by virtue of this fact can ice float on the surface of water. Between zero and four degrees, water shows an exception to the general rule that we find when temperature increases, namely, that bodies become less and less dense as they are heated. This range of four degrees, in which water expands as the temperature is lowered, is very instructive. What do we learn from this range? We learn that the water begins to struggle. As ice it is a solid body with a solid's inner relationships, a kind of individuality. Now it is to pass over selflessly into relation to the whole earth, but it doesn't pass into this state of selflessness easily. It struggles against the transition into an entirely different sphere. It is necessary to consider such things, for then we begin to understand why, under certain conditions, the temperature rise as determined by a thermometer disappears, say at the melting-point or boiling-point. It disappears just as our bodily reality disappears when we rise to the realm of Imagination.[31] Just as it is possible for our bodily activity to pass over into the spiritual when we enter the imaginative realm, so we can find a path leading from the external and visible in the realm of heat to the underlying phenomena indicated by our thermometer when the temperature rise we are measuring with it disappears before our eyes.

## First steps of a new method

You know that solid bodies such as most metals and other mineral bodies do not occur in an indefinite form but in very definite shapes that we call crystals. We can say that under ordinary circumstances as they exist on the earth, solids occur in very definite shapes or crystal forms. This naturally leads us to turn our attention to these forms and to try to puzzle out how these crystals originate. What forces underlie crystal formation? In order to gain some insight into these matters, it will be necessary for us to consider how solid bodies behave that are on the surface of the earth but not connected directly with the mass of the earth.

You know that when we are holding a solid in our hand and let go of it, it falls to the earth. In physics this is usually explained as follows: the earth attracts solid bodies, exerts a force on them; under the influence of this force—the gravitational force—the body falls to the earth. When we have a fluid and cool it so that it solidifies, it may also form definite crystal formations. The question now is: what is the relationship between the force acting on all solids—gravitation—to those forces that must be present and active in a definite way for the solid body to manifest in crystalline shapes? You might well think that gravity as such, through whose agency a body falls to the earth (if we even want to speak of forces such as gravity), that this gravitational force had nothing to do with the building of crystal form. For gravity affects all crystal forms alike. No matter what outer form an object may have, it is subject to gravity. We find, when we have a number of solids in a row and take away their support, that they all fall to earth in parallel lines. This fall may be represented in somewhat the following way [*see Fig. 10*].

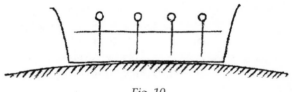

Fig. 10

We can say that whatever form a solid may have, it falls along a line perpendicular to the surface of the earth. When we draw the perpendicular to these parallel lines of fall, we obtain a surface parallel to the earth's surface. By drawing all possible perpendiculars to the lines of fall, we will obtain a complete surface parallel to the earth's surface. This is at first an imagined surface. We may now ask the question: where is this surface real? It is real in fluid bodies. A liquid placed in a vessel shows as a real liquid surface what I have assumed to be produced by drawing perpendiculars to the line of fall of solids.

What is really involved here and what does it mean? What we have just discovered is something of tremendous significance. For imagine the following. Suppose someone were trying to explain the nature of the surface of a liquid and stated it in this way. Every minute portion of the liquid has the tendency to fall to the earth. Since the forces of the liquid itself hinder this,[32] the liquid surface is formed. It is the fluidity that causes the surface to form. Picture to yourselves now the initial position of the solid bodies that you let fall, and then nature herself draws for you what you have drawn for purposes of this explanation. You must add the surface level in your thinking. I said previously that the surface level is to be thought of in its relation to solids as at right angles to their line of fall. When you think through these thoughts, you will make the curious discovery that what you normally do in order to relate thoughts to liquid is done for you by a number of solid bodies. The body of a lower state of aggregation — the solid body in its behaviour towards the earth's surface — reveals to us as if in a picture what is actually present in the liquid, what is materially there in the liquid preventing the realization of this line as the drop line. This becomes pictorial if I consider the solid body in its entire relation to the earth.

Think what this enables us to do. When I draw the lines of fall and the surface formed under the impression of a system of falling bodies, then I have a picture of gravitational activity. This is a direct image of liquid matter.

We can proceed further. When we leave water at any temperature sufficiently long, it dries up. Water is always evaporating. The conditions under which it forms a liquid surface are only relative. It must be confined all around except on the liquid surface. It evaporates continuously, and does so more rapidly in a vacuum. If we draw lines showing the direction in which the water is tending, their direction must indicate the movement of the water particles when it evaporates. When I actually draw these lines, however, I get nothing more or less than an image of a gas that is enclosed all around and is striving to escape in every direction [*see Fig. 11*].

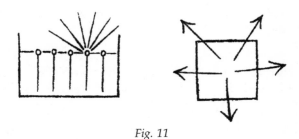

Fig. 11

There is a certain tendency on the surface of water which, when I picture it for explanatory purposes, represents a gas set free and spreading itself in all directions. We can state the proposition again in this way: what we observe in water as a force is actually a picture of what gas is as a material reality.

A curious fact becomes apparent here. If we look at fluidity correctly from a certain point of view, we discover in this state a picture of the gaseous state of aggregation.[33] If we properly observe solids, we perceive pictures of the fluid state of aggregation. In every subsequent downwards state, pictures of the preceding state develop. Let us illustrate this by going from below upwards. We can say that in solids we discover the images of the fluid state, in the fluid we discover the images of the gaseous, in the gaseous we discover images of warmth. We will deal especially with this in the next lecture. For now I will say only that we have attempted to find the transition in thought from gas to warmth. It will become clearer next time.

Now when we have pursued this path of thinking further, that:

In solids we have images of the fluid state
In fluids we have images of the gaseous state
In the gaseous we have images of the warmth state

then we will have made an important step. We have advanced to the point where we have a picture in the gaseous state which is accessible to human observation, of heat manifestations and even of the real nature of heat itself. We then gain the possibility, if we are seeking images of heat in the gaseous state in the right way, to explain the nature of heat even though we are obliged to admit that it is an unknown entity to us at the outset. We must look for images of the nature of heat in bodies in the gaseous condition.

But we must do this in the right way. When the various phenomena that we have described so far are handled as physics usually handles them, we get nowhere. If, however, we consider correctly the things that are revealed to us by bodies under the influence of heat and pressure, we will see how we actually come to stand before what the gases can reveal to us — the real nature of heat.

In cooling, the entity which is heat penetrates further into liquid and solid states. We have to pursue the nature of this heat entity also in these states, although we can do it best in the gaseous condition where it is more evident. We must see whether in the fluid and solid states the nature of heat undergoes a special change within itself. Through this distinction between its manifestation in the gaseous realm, where it reveals itself to us in pictures, and its manifestation in fluids and solids, we must arrive at the real nature of heat itself.

# 9. What is Warmth?

*Here Rudolf Steiner attempts to clarify the essential nature of heat or warmth from a new direction. It is not merely a form of energy, or the motion of atoms and molecules, although it is connected with these phenomena. In the 'Warmth Course', the concept gradually emerges of warmth occupying a unique position between the physical and spiritual realms, and with the opposite character to gravity.*

## Not just energy

If we observe warmth phenomena in a solid body, what we have, fundamentally, is the solid body on the one hand and on the other a process in the realm of warmth or heat. We see before our eyes, as it were, phenomena within the realm of warmth that we also see taking place in a gas. From this we can conclude — or rather simply state, since we are only describing what is obvious — that if we wish to approach the true being of warmth we must seek as far as possible to penetrate into the gaseous realm, into gaseous bodies. And in what goes on in gases we will see images of what takes place within the warmth realm. By a manifestation of certain phenomena in gases, therefore, nature conjures up before our eyes, as it were, pictures of heat processes. Notice now that we are being led very far from the modern method of observation as practised in natural science generally, not merely physics. Let us ask ourselves where the modern method really leads us ultimately. I have here a work by Eduard von Hartmann, in which he treats the field of modern physics from his point of view. Here is a man who has created for himself a broad horizon entirely out of the spirit of the times and who we may say is therefore in a position to speak about physics as a philosopher. Now it is interesting to see how such a man deals with physics, speaking entirely in the modern spirit. He begins the very first chapter as follows: 'Physics is the study of transformations and movements of energy and of its separation into factors and its resultants.' Having said this, he must naturally add a further statement:

> Physics is the study of the movements and transformations of energy (force) and of its resolution into factors and its resultants. The validity of this definition is not dependent on how we consider energy. It does not rest on our considering it as something final, ultimate, nor on our looking upon it as really a product of this or that view of the constitution of matter. It presupposes that all perception and sensation point back to energy, that energy can change place and form and that, in accordance with its concepts, it can be analysed.

Now what does it mean when one speaks in such a way? It means that an attempt is made to define what confronts us physically without needing to enter into its real nature — a definition which, by its very nature, makes it unnecessary to enter into the real nature of what is defined. A certain concept of energy is formed, to the effect that everything which meets us from without, physically, is only a transformation of this energy concept. That is to say, everything essential is discarded from one's concepts; and then the certain belief is promoted that, even if nothing else can be grasped, at least we have created secure definitions. But this sort of thing has found its way into our physical concepts to a most unfortunate extent. So completely has it become a part of our view of physics that it is almost impossible today for us to do experiments that reveal reality to us. All the laboratories that we depend upon to do physical research are completely given over to working out the theoretical views of modern physics. We cannot easily use what we have in the way of tools in order to penetrate physically into the nature of things.

## Negative gravity and negative form

We have seen that the fundamental property of solid bodies is the possession of form. The solid body emancipates itself, so to speak, from what is active in creating the form of a liquid, active at least relatively if the liquid does not evaporate in the course of time. The solid body thus has its form somehow. Liquids must be enclosed in a vessel, and in order to form a liquid surface such as we find everywhere on the surface of a solid they require the forces of the entire earth. We have considered this, and it now requires us to make the following statement: When we consider the liquids of the whole earth in their totality, we are obliged to consider them as related to the one body of the earth in its totality. Only the solids emancipate themselves from this relationship to the earth; they take on an individuality, assume their own form. If we now take the method by which ordinary physics represents things and bring it to bear on what is called gravity, on what causes the formation of the liquid surface, then we must do it in the following way. If we are to stick to what is observable, we must in some way introduce into indivi-dualized solid bodies the thing that is essential in this horizontal liquid surface. In some way or other, we must conceive of what is active in forming the surface of a liquid, and which is thought of under the heading of gravity, as being *within* solids; they then individualize gravity in a certain way. Thus we see that solids take gravity up into themselves.

On the other hand, we see that at the moment of evap-oration the formation of a liquid surface ceases. Gas does not form a surface. If we wish to give form to a gas, to limit the space it occupies, we must do so by placing it in a vessel closed on all sides. In passing from liquid to gas, we find that surface-formation ceases. We see dissipated this last remainder of the earth-induced tendency to surface formation as shown by the liquid. And we see also that all gases are grouped together in a unity, as illustrated by the fact that they all have the same coefficient of expansion—

gases as a whole represent material emancipated from the earth.

Now place these thoughts vividly before you. You find that as human beings on the solid earth—as a carbon-aceous organism therefore—you are among the phenom-ena produced by the solids of the earth. The phenomena produced by solids are ruled by gravity which, as stated, manifests itself everywhere. As earthly human beings you have solids around you that have in some way taken up gravity to create their form. But consider the phenomena manifested by the solids in the case I spoke of before, where you mentally add a surface to what is occurring. In this phenomenon you have a kind of continuum, some-thing you can think of as a sort of invisible fluidity spread out everywhere. Thus the solids of the earth, in so far as they are free to move, represent a fluid in the sum of their manifestations. They constitute something similar to what is manifested in material fluidity. We can therefore say that since we are placed on the solid earth we perceive what is surface-forming in water and call it gravity.

Imagine now that as human beings we were able to live on a fluid cosmic body, being organized in such a way that we could exist on such a body. Then we would have to live above the surface level of this liquid[34] and we would have the same relationship to the gaseous, striving outward in all directions, that we now have to the fluid. This means nothing more nor less than that we would not perceive gravity. To speak of gravity would cease to have a meaning. Gravity is thus perceived only by those beings that live on a solid planet, and only those bodies are subject to it that are solid. Beings who could live on a fluid planet would know nothing of gravity. It would not be possible to speak of such a thing. And beings who lived on a gaseous planet would regard as normal something that would be the opposite of gravity, a striving in all directions away from the centre. If I may express myself somewhat paradoxically, I might say that beings dwelling on a gaseous planet instead of seeing bodies falling towards the planet would see them always flying off. If we are really to find our way, we must think in truly physical terms and not merely in mathematical terms. Then we can state things as follows. Gravity begins when we find ourselves on a solid planet. In passing from the solid to the gaseous planet, we go through a kind of zero-point and come to an opposite condition to that on the solid planet, to a manifestation of forces in space that may be considered negative in respect to gravity [*Fig. 12*].

You see, therefore, that as we pass through these material states, we actually come to a zero-point in spatial

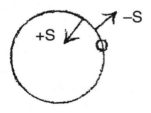

*Fig. 12*

terms, to a zero-sphere in spatial terms. For this reason we have to consider gravity as something quite relative.

But when we conduct warmth to a gas (we did this experiment for you), this warmth which always raises the diffusing tendency in the gas shows you again the picture I am trying to present [*see Fig. 11*]. Is it not the case that what is active in the gas really lies on the far side of this zero-sphere to which gravity tends? Is it not possible for us to think the matter through further, still closely observing the actual phenomena, and say that in going from a solid to a gaseous planet we pass through a zero-point? Below we have gravity; above, for physical thinking, this gravity changes into its opposite, into a negative gravity. Indeed we find this, we do not simply have to think it. The being of warmth does the same thing that a negative gravity would do. Certainly, we have not completely attained our goal, but we have reached a point where we can comprehend the being of warmth in a relative fashion to the extent that the matter may be stated thus: the being of warmth manifests exactly like the negation of gravity, like negative gravity. When one therefore deals with physical formulae involving gravity and sets a negative sign in front of the symbol representing gravity, it is necessary to think of the magnitude in question not as a gravity quantity nor as a sequential scale of gravity but as a warmth quantity, a sequential scale of warmth. You can see that in this way we are able to suffuse mathematics with life. The formulae as they are given may be looked upon as representing a gravitational system, a purely mechanical system. If we set negative signs in front of the quantities, then we are obliged to consider as warmth what formerly represented gravity. And you realize from this that we must grasp these things concretely if we are to arrive at real results. We see that in passing from the solid to the fluid we go through a condition in which form is dissolved. The form is lost. When I dissolve a crystal or melt it, it loses the form that it previously had. In passing into fluidity it takes on the form that is imposed upon it by virtue of the fact that it comes under the general influence of the earth. The earth gives it a liquid surface, and I must put this liquid into a vessel if I am to preserve it.

Now let us consider another general phenomenon which we will approach more concretely later. If a liquid is divided into sufficiently small particles, drops are formed which take on the spherical shape. Fluids have the possibility, when they are subdivided finely enough, of emancipating themselves from the general gravitational field and of manifesting in this special case what otherwise comes to light in solids as polyhedral form, as crystalline shape. In the case of fluids, however, the peculiarity is that they all take on the form of the sphere. If I now consider the spherical form, I may regard it as the synthesis of all polyhedral shapes, of all crystal forms.

When I now pass from fluid to gas, I have the diffusion, the dissolution of the spherical form, but in this case directed outwards. And now we come to a rather difficult idea. Imagine to yourselves that you are observing some simple form, say a tetrahedron, and that you wished to turn it inside out as you might a glove. You will then realize that in going through this process of turning inside out it is necessary to pass through the sphere, and then the negative body appears, for which all corresponding relations are negative. As the tetrahedron is put through this transformation, you must imagine this negative body in such a way that the entire space outside the tetrahedron is filled, is gaseous. Within this filled outside space you must imagine a tetrahedral hole. There it is hollow [*Fig. 13*]. You must then make the quantities related to the tetrahedron negative. Then you have formed the negative, the opened-up tetrahedron, in place of the one normally filled with matter. But the intermediate condition between the positive and the negative tetrahedron is the sphere. Every polyhedric body goes over into its negative only by passing through the sphere, as through a zero-point, a zero-sphere.

Now let us follow this concretely in the case of actual bodies. You have the solid bodies with definite forms. They go through the fluid form, that is the sphere, and

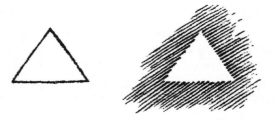

*Fig. 13*

become a gas. If we wish to regard the gas in the right way we must look upon it as form, but as negative form. We reach a type of form here that we can comprehend only by passing through the zero-sphere into the negative. That is to say, when we go over to the gaseous — the image of warmth phenomena — we do not enter into the region of the formless, we only enter a region more difficult to comprehend than the one in which we live ordinarily, where form is positive and not negative. But here we see that any body in which the fluid state comes into consideration is in an intermediate condition. It is in the state between the formed and what we call the 'formless', or that of negative form.

## Warmth between the physical and the spiritual

I said to you that we have, to begin with, the realm of solids. Solids are the bodies which manifest in definite forms. We have, touching on the realm of the solids, as it were, the realm of fluids. Form is dissolved, disappears, when solids become liquids. In the gaseous bodies we have a striving in all directions, a complete formlessness — negative form. Now how does this negative form manifest itself? If we look in an unbiased way at gaseous or aeriform bodies we can see in them what may be considered as corresponding to the entity elsewhere manifested as form. I have already called your attention to the realm of acoustics, the tone world. In the gaseous state, as you know, the manifestation of tone arises through condensations and rarefactions. But when we change the temperature we also have to do with condensation and rarefaction in the body of the gas as a whole. Thus if we pass over the liquid state and seek to find in the gas what corresponds to form in the solid, we must look for it in condensation and rarefaction. In the solid we have a definite form; in the gas, condensation and rarefaction.

And now we pass to the realm adjacent to the gaseous. Just as the fluid realm borders on the solid, and just as we know how the solid creates an image of the fluid and the fluid creates an image of the gaseous realm, so gas creates an image of the realm we must conceive as lying next to the gaseous, i.e. the realm of heat. The realm lying next beyond heat we will have to postulate for the time being, calling it the x region.

| $x$ | Becoming material — Becoming spiritual | |
|-----|------------------|------------------|
| Heat | | |
| Gas | Negative form | Condensation — Rarefaction |
| Fluid | | |
| Solids | Form | |

If now I seek to proceed further, at first merely through analogy, I must look in this x region for something corresponding to but beyond condensation and rarefaction (this will be verified in our subsequent considerations). I must look for something like condensation and rarefaction there in the x region, passing beyond heat just as we passed beyond the lower, fluid state. If you begin with a solid, closed form, then imagine it becoming gaseous, by this process simply having changed its original form into another fluid form manifesting as rarefaction and condensation. If you then think of the condensation and rarefaction as heightened in degree, what is the result? As long as condensation and rarefaction are present, obviously matter is still there. But now, if you rarefy further and further you finally pass entirely out of the realm of the material. And this continuation we have spoken of must, if we are to be consistent, be expressed as a boundary between the material and the spiritual. When you pass beyond the heat realm into the x realm, you enter a region where you are obliged to speak of the condition in a certain way. Holding in mind this transition from solid to fluid, and the condensation and rarefaction in gases, you pass to a region of materiality and non-materiality. You cannot do other than speak of the region of materiality and non-materiality. This means that when we pass through the heat realm we actually enter a realm which is in a sense a consistent continuation of what we have observed in the realms beneath it. Solids oppose heat—heat cannot come to complete expression in them. Fluids are more susceptible to the intentions of heat. Gas completely follows the intention of heat. In gases there is a thorough manifestation of heat—it plays through them without hindrance, doing with the gas what it wants. In its material processes, gas is an image of heat. I can put it like this: the gas in its material behaviour is essentially similar to the heat entity. The degree of similarity between matter and heat becomes greater and greater as I pass from solids through fluids to gases. This means that liquefaction and evaporation of matter signify that matter becomes more similar to heat. Passage through the heat realm, however, where matter becomes, so to speak, identical to heat, leads to a condition where matter itself ceases to be. Heat thus stands between two strongly contrasted regions, essentially different from each other — the spiritual realm and the material realm. Between these two stands the realm of heat. This transition zone is really somewhat difficult for us. We have on the one hand to climb to a region where things appear more and more spiritualized, and on the other hand to descend into what appears more and more material. Infinite extension upwards appears on the one hand and infinite extension downwards on the other [*indicated by arrows in the diagram above*].

Now let us use another analogy. I am presenting it to you today because, through a general view of individual natural phenomena we can develop a sound natural science. Perhaps it will be useful to make an array of these facts while we are considering them. If you observe the usual spectrum, you have red, orange, yellow, green, blue, indigo, violet.

Infrared, red, orange, yellow, blue, indigo, violet, ultraviolet

You have a band of colours following each other in a series of approximately seven nuances. But you also know that the spectrum does not break off at each end. If we follow it further beyond the red we come to a region where there is more and more heat, and finally we arrive at a region where there is no light but only heat, the infrared region.

On the other side of the spectrum, the violet, we also come to a region where we no longer have light. We come to the ultraviolet where only chemical action is manifested or, in other words, effects that manifest themselves in matter.

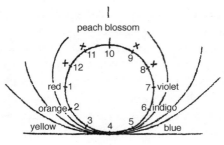

*Fig. 14*

You also know, however, that according to Goethe's colour theory, this series of colours can be bent round into a circle and arranged in such a way that one sees not only the light from which the spectrum is formed but also the darkness from which it is formed [*see Fig. 14*]. In this other spectrum the colour in the middle is not green but peach-blossom colour, and the other colours proceed from this.

When I observe darkness I obtain the negative spectrum. And if I place the spectra together, I have twelve colours that may be definitely arranged in a circle: red, orange, yellow, green, blue, indigo, violet; on this side the violet becomes ever more and more similar to the peach-blossom and there are two nuances between; on the other side there are two nuances between peach-blossom and red. You have in all, if I may use the expression, twelve colour conditions. You can see from this that what is usually described as occurring in a linear spectrum can, by suitable means, also be achieved by this circle of colour. We can make it larger and larger, stretching out the five upper colours until they finally disappear. The lower arc becomes practically a straight line, and I obtain the ordinary spectrum of colours, having brought about the disappearance of the upper five colours.

So much for the colours. Could it not also be that here [*see the chart at the beginning of this section, p. 142*] passing into infinity is somewhat similar to what I have done here to the spectrum? Suppose I ask what happens if what apparently goes off into infinity is made into a circle and returns on itself. Am I not perhaps dealing here with another kind of spectrum, one that, on the one hand, encompasses the conditions extending from heat to matter, but which, on the other hand, I can close up into a circle as I did in introducing peach-blossom into the colour spectrum?

# 10. The Scale of Nature

*Having reinstated the elements in their rightful place, and discovered the pivotal role of warmth or heat as a region mediating the physical and the spiritual, Rudolf Steiner applies the method of deducing the essential characteristic of each realm from certain properties of the preceding realm, to deduce the existence of supersensory regions beyond warmth. He also arrives at a region below the solid, preceding the realm of forms. The regions above warmth are characterized as non-spatial, with the opposite dynamic to physical forces. The term 'suction' is used to describe them, not in the earthly sense of a lower pressure, but in the sense of being polar opposites. The term 'peripheral' was used in the* Light Course *(see Chapter 6).*

*Projective geometry (not mentioned in either the* Light Course *or the* Warmth Course, *but in* Anthroposophy and Science — *see 'Sources' p. 233) has proved a fruitful way of giving these concepts mathematical form. Steiner pointed to the importance of this hitherto little-known branch of mathematics as a way of grasping the inner essence of form. Recent research has indicated that heat and the realms above it can be understood in terms of 'counterspace', the polar opposite of space.*[35]

Once more I wish to draw out the essential thing for you. We proceed from the realm of solids and find a common property at first manifesting in form. We then pass through the intermediate state of the fluid, showing form only to the extent of making a liquid surface for itself. Then we reach the gaseous state where the property corresponding to form in the realm of solids manifests itself as condensation and rarefaction, even in a realm formless in nature. We then come to the region bordering on the gaseous, the heat region, which again, like the fluid, is an intermediate region. And finally we arrive at our x realm. We saw that by pursuing our thought further we have to postulate in x processes of materialization and dematerialization. It is not difficult, then, to see that we can obviously go beyond x to a y and a z realm, just as in the light spectrum, for instance, we move from green to blue, from blue to violet, and then into ultraviolet.

| | |
|---|---|
| z | |
| y | |
| x | Materialization — Dematerialization |
| Heat region | |
| Gaseous bodies | Rarefaction — Condensation |
| Fluid bodies | |
| Solid bodies | Form |
| U | |

And now it is a question of studying the mutual relations between these different regions. In each one we see appearing what I might call definitely characteristic phenomena. In the lowest realm we see an enclosed form; in the gaseous realm we see a fluctuating form, so to speak, in condensation and rarefaction. This accompanies — and I am now speaking precisely — this accompanies the tone entity under certain conditions.[36] When we pass through the warmth realm into the x realm, we see materialization and dematerialization. The question now arising is this: how does one realm work into another?

I have already called your attention to the fact that when we speak of gas, the processes taking place in the gaseous realm present a kind of picture of what goes on in the realm of heat. We can say, therefore, that in gas we find a picture of what goes on in the heat realm. We have to consider that gas and heat mutually interpenetrate each other so that gas in its material form does what the heat wishes; in the processes that take place within a gas-filled space we can see an image of warmth. What heat wishes grasps hold of gas in its expansion in space. What is really taking place in the realm of heat expresses itself in the gas in the interpenetration of the two realms. Furthermore we can say that fluidity shows a relationship to the gaseous similar to that between the gaseous realm and heat. And solids show the same sort of relationship to fluids as fluids do to gases and gases do to heat.

What comes about, then, in the realm of solids? In this realm forms appear, definite forms, forms circumscribed within themselves. These circumscribed forms are pictures, as it were, of what is really active in fluids. Now we can pass here to a realm U, below the solid, whose existence we will at first merely postulate hypothetically. And let us try to create concepts in order then to see whether these concepts are somehow applicable in the realm of outwardly perceptible phenomena. By extending our thinking, which you can feel is rooted in reality, we can create concepts that we hope will then lead us, because they were gained from reality, a bit further into reality.

What must take place if there is to be such a reality as the U realm? In this realm there must be contained an image or picture of what in solids is a manifest fact. In a way that corresponds to the other realms, the U realm must give us a *picture* of the realm of solids. In the realm of solids, we have forms everywhere, forms that are shaped out of their own intrinsic being, or at least out of their relation to the world. Forms come into being mutually interrelated.

Let us go back for a moment to the fluid state. There, through the outwardly enclosed surface level of the fluid, we have a body showing its relation to the entire earth. In gravity, therefore, we have to recognize a force akin to the creation of form in solids. In the U realm we must find something that happens similar to the form-building in the world of solids, if we are to pursue our thinking in accordance with reality; this would parallel the way solids prefigure or create an image of the fluid world. In other words, in the U realm we must be able to see an action that foreshadows the various formations of the solid world. We must in some way be able to see this activity. We must see how, under the influence of different forms related to each other, something else arises. Something must come into reality that arises under the influence of the varying forms in the solid world. Today we really have only the beginning of such an insight. Suppose you take a substance such as tourmaline, which carries a form principle within itself. You then let this formed tourmaline act in such a way that form can act on form. I am referring to the inner formative tendency. You can do this by allowing light to shine through a pair of tourmaline crystals. At one moment you can see through them, and then the field of vision darkens. You can bring this about simply by turning one crystal in relation to the other. You have thus brought their form-creating force into a different relationship. This phenomenon is inwardly related to the apparent passage of light through systems of solids differing in form, showing us the so-called polarization figures.[37] These polarization phenomena always appear when one form influences another. Here before our eyes we have the noteworthy fact that we look through the solid realm into another realm that is related to the solid, just as the solid is to the liquid. Let us ask ourselves now, how does it come about that, under the influence of the form-building force, something appears in the U realm that creates form in the realm of solids, just as gravity forms only the surface in the fluid realm? To this we must reply that this happens when we observe the so-called polarization figures that lie in a realm to be found beneath that of the solids. We are actually looking into a realm that underlies the world of solids. But we see something else also. We might look for a long time into such a system of solids, and the different forces might be acting there upon each other in the most varied ways, but we would see nothing if there were not something else playing through these solids as the U realm permeates the world of solids. Light, for instance, penetrates in and makes this mutual interworking of the form-building forces visible for us.

What I have been describing here is treated by the physics of the nineteenth century in such a way that the light itself is said to give rise to the phenomenon, while in reality the light only makes the phenomenon *visible*. If one wishes to comprehend these polarization figures, one must seek for an entirely different source than the light itself. What is taking place has nothing to do with the light as such. The light simply penetrates this U realm and makes visible what is going on there, what is taking place there as a foreshadowing of the solid form. Thus we can say that we have to do with an interpenetration of different realms that we have simply revealed here. In reality we are dealing with an interpenetration of different realms.


## Spatial and non-spatial, pressure and suction

At this point I would like to build a bridge, as it were, between the deliberations in this Course and those in the Light Course. Today we will study the light spectrum, as it is called, and its relation to the heat and chemical effects that come to us with the light. The simplest way for us to bring before our minds what we are to deal with is first to make a spectrum and learn what we can from the behaviour of its various components. We will therefore make a spectrum by throwing light through this opening—you can see it here. [*The room was darkened and the spectrum shown.*] It can be seen on this screen. You can see that we have something hanging here in the red portion of the spectrum. We can observe something by means of this instrument. First we wish to show you how heat effects arise especially in the red portion of the spectrum. We can observe these effects through this expanding action of the energy cylinder on the air contained in the instrument, which expanding action in turn pushes the alcohol down on this side and up on this one. This depression of the alcohol column shows us that there is a considerable heat effect in this part of the spectrum. It would also be interesting to show that when the spectrum is moved so as to bring the instrument into the blue-violet portion, the heat effect is not noticeable. This heat effect is essentially characteristic of the red portion.

And now, having shown the occurrence of heat effects in the red-yellow portion of the spectrum by means of the alcohol column, let us show the chemical activity of the blue-violet end. We do this by allowing the blue portion to fall on a substance that you can see is brought into a state of phosphorescence. From the previous Course you know that this is a form of chemical activity. Thus you see an essential difference between the portion of the spectrum that disappears into the unknown on this side and the portion that disappears on this other side; you see how the substance glows under the influence of the chemical rays, as they are called.

Moreover, we could arrange matters in such a way that the middle portion of the spectrum, the real light portion, would be cut out. We could not do this with absolute precision, but we could make the middle portion approximately dark by simply placing in the light's path a solution of iodine in carbon disulphide. This solution has the property of stopping the light. It is possible to demonstrate the chemical effect on one side and the heat effect on the other side of this dark band. Unfortunately we cannot carry out this experiment completely but only mention it in passing. If I placed an alum solution in the path of the light, the heat effect would cease, and you would see that the alcohol column was no longer displaced because the alum, or the solution of alum, to speak precisely, would hinder the passage of the heat effect. Having placed alum in the light's path, you would soon see the column equalize, because the heat would no longer be present. We would have here a cold spectrum.

It is very interesting that a solution of aesculin[38] placed in the path of the light source will cut out the chemical effect of the spectrum. Unfortunately we could not get this substance. The heat effect and the light effect would remain, but the chemical effect would cease.

Now let us place in the light's path the solution of iodine

in carbon disulphide; the middle portion of the spectrum disappears. You see clearly the red portion — it would not be there if the experiment were an entire success — and the violet portion, but the middle portion is dark. We have succeeded partly in our attempt to eliminate the bright portion of the spectrum. By carrying out the experiment in a complete way, as certain investigators have done (for instance, Dreher, 50 years ago), the two bright portions you see here can be done away with completely. Then the temperature-elevating effect may be demonstrated on the red side, and on the other side phosphorescence will show the presence of the chemically active rays.[39] This has not yet been fully demonstrated, and it is of very great importance. It shows us how what we think of as active in the spectrum can be conceived in its general cosmic relations.

In the Course I gave here previously, I showed how a powerful magnet works on relations within the spectrum. The force emanating from the magnet alters certain lines, changes the formation of the spectrum itself. It is only necessary for a person to extend the thought prompted by this in order to enter the physical processes in his thinking. You know from what we have already said that there is really a complete spectrum, a collection of all possible colours yielding a spectrum of twelve colours — then we have a circular spectrum instead of the spectrum spread out in one dimension of space. In the circular spectrum we have green here, peach-blossom here, here violet, and here red, with the other shades between — twelve shades clearly distinguishable from one another [*Fig. 15*, over].

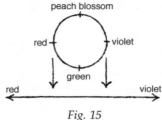

*Fig. 15*

The fact is that under earthly conditions such a spectrum can be presented only as an image. When we are dealing with this spectrum in the domain of earthly life, we can do so only by means of a picture. The spectrum we actually get is the well-known linear one extending as a straight line from red through green to blue and violet—thus we obtain a spectrum formed from the circular one, as I have often said, by making the circle larger and larger, so that the peach-blossom disappears, violet shades off into infinity on one side, and red shades off on the other, with green in the middle.

We may ask how this partial spectrum, this fragmentary colour band, arises from the complete series of colour, the twelve-colour series that must be possible. Imagine that you have the complete circular spectrum, and suppose forces act on it to make the circle larger and larger and finally to break it at this point [*see Fig. 15*]. Then, when the circle has broken, the action of these forces would make a straight line of the circle, a line extending apparently into infinity in each direction.

When we come upon this linear spectrum under our earthly conditions, we feel obliged to ask the question: how is this able to arise? It can arise only by means of the seven known colours separating out. They are, as it were, cut out of the complete spectrum by the forces that work into it. But we have already come upon these forces in the earthly realm. We found them when we turned our attention to the forces of form. This too is a formative activity: the circular form is turned into the straight-line form. It is a form that we meet with here. And considering the fact that the inner structure of the spectrum is altered by magnetic forces, it becomes quite evident that the forces making our spectrum possible are everywhere active here. This being the case, we have to assume that our spectrum, which we consider a primary thing, has already working within it certain forces. Not only must we consider light variation in our ordinary spectrum, but we have to think of this ordinary spectrum as including forces that render it necessary to represent the spectrum by a straight line.

We must link up this line of thought with another, which comes to us when we go through the series — as we have done frequently before:

| Materializing, dematerializing; dark, light | |
|---|---|
| Heat | |
| Rarefying, densifying | |
| Liquid | |
| Solid form | |

We pass from solid forms, through fluids, to condensation and rarefaction, i.e., gases, to heat, and then to that state we have called x, where we have materialization and dematerialization. Here we meet a higher enhancement of condensation and rarefaction, beyond the heat condition, just as condensation and rarefaction proper constitute an enhancement of forms becoming fluid. When form itself becomes fluid, when we have a changing form in a gaseous body, that is an enhancement of a definite form. And what occurs here? An enhancement of condensation and rarefaction. Keep this definitely in mind, that we enter a realm where we have an enhancement of condensation and rarefaction.

What do we mean by an 'enhancement of rarefaction'? Well, matter itself informs us what happens to it when it becomes more and more rarefied, it tells us what matter is struggling with. When I make matter more and more dense, a light placed behind the matter no longer shines through. When the matter becomes more and more rarefied, the light does pass through. When matter is rarefied enough, I finally come to a point where I obtain brightness as such. Therefore, what I bring into my understanding here in the material realm is empirically found to be the genesis of brightness. Dematerialization will appear to me as brightness; materialization will always appear to me as darkness. I thus have to think of brightness or luminosity as an enhancement of the condition of rarefaction. And darkening has to be thought of as a condensation not yet intense enough to produce matter, but of such an intensity as to be just on the verge of becoming material.

Now you see how I can place the realm of light above the heat realm and how the heat realm is related to that of light in an entirely natural way. But when you recollect how a given realm always gives a sort of picture of the realm immediately above it, then you must look in the nature of heat for something that foreshadows, as it were, the conditions of luminosity and darkening. In the heat that appeared at one end of the spectrum, we must find something that provides a picture of lightening and darkening. Keep in mind that we do not find only the upper condition in the lower, but also the lower condition always in the upper as well. When I have a solid body, it can be altogether in the fluid realm with its solidity.[40] What gives it form may extend over into the next realm, which is not characterized by form. If I wish to encompass reality with my concepts I must be clear that there is a mutual interpenetration of qualities of reality. This principle takes on a particular form, however, in the realm of heat; dematerialization works down into heat from above [*see arrow in the diagram*], while from the lower side the tendency to materialization works up into the heat realm.

Thus you see that I draw near to the nature of heat when I see in it on one side a striving for dematerialization and on the other a striving for materialization. If I wish to grasp the nature of heat, I can do so only by conceiving within it a life, a living, weaving activity, manifesting itself everywhere as a tendency to materialization penetrated by a tendency to dematerialization. What a gulf exists between this conception of heat based on reality and the nature of heat as outlined by Clausius' so-called mechanical theory of heat! In the Clausius theory we have atoms or molecules in a closed space, little spheres moving in all directions, colliding with each other and with the walls of the vessel in a purely external movement [*Fig. 16*].

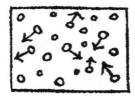

*Fig. 16*

And it is actually stated that heat consists of this chaotic movement, of this chaotic collision of material particles with each other and with the walls of the vessel. A great controversy arose as to whether the particles were elastic or non-elastic. This is of importance only in respect to whether the phenomena can be better explained on the assumption of elasticity or on the assumption that the particles are hard, non-elastic bodies. This has given rise to the conviction that heat is purely motion in space and so it is stated: *Heat is motion.*

Actually, we must agree that heat is motion, but in an entirely different sense. It is motion in which, wherever heat is manifested in space, there is a tendency to create material existence and to let material existence disappear again. It is no wonder that we also need heat in our organism. We need heat in our organism simply to continually change what exists spatially into what has no existence in space. If I simply walk through space, what my *will* is producing shapes the space. When I think about it, something completely non-spatial is present. What makes it possible for me as a human organization to be outwardly integrated into the form relationships of the earth? When I walk over the earth, I change the entire earthly form. I change its form continually. What makes it possible that I am connected with everything on the earth and that I can form non-spatial ideas within myself, as observer of what is manifested in space? What makes this possible is that my existence is enacted in the medium of heat and is thus continually enabled to transform material effects, that is spatial effects, into immaterial effects that no longer partake of spatial nature. In myself, in fact, I experience what heat is in reality: intensive motion, motion that continually alternates between the realm of pressure effects and that of suction effects.

Assume that you have here [*Fig. 17*] the border between pressure and suction effects.

The effects of pressure run their course in space, but the suction effects do not,[41] as such, act in space — they operate outside space. For my thoughts, based on the effects of suction, take their course outside of space. Here on one side of this line, I have the non-spatial. And now when I conceive of what takes place neither in the realms of

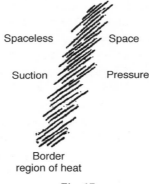

Spaceless    Space

Suction    Pressure

Border
region of heat

*Fig. 17*

pressure nor suction but on the border between the two, I am dealing with the things that take place in the realm of heat. I have a continual searching for equilibrium between pressure effects of a material sort and suction effects of a spiritual sort. It is quite remarkable that certain physicists have had these things right under their noses but refuse to consider them. Planck, the Berlin physicist, has made the following striking statement: If we wish to arrive at a concept of what is called ether nowadays, the first requirement is to follow the only path open to us, in view of the knowledge of modern physics, and consider the ether non-material. This from the Berlin physicist, Planck! The ether, therefore, is not to be considered as a material substance.

But now, what we find beyond the heat region, the realm in which the effects of light take place, we consider so little allied to the material that we are assuming the pressure effects — characteristic of matter — to be completely absent, and only suction effects active there. This means that we leave the realm of ponderable matter and enter a realm naturally active everywhere but manifesting itself in a way diametrically opposed to the realm of the material. We must picture only suction effects active there, emanating from every point of space, while material things obviously manifest through pressure effects. Thus, indeed, we come to an immediate concept of the nature of heat as intensive motion, as an alternation between pressure and suction effects, but in such a way that we do not have, on the one hand, suction spatially manifested and, on the other hand, pressure spatially manifested. If we wish to comprehend the nature of heat, we must entirely leave the material world and with it three-dimensional space. If the physicist expresses certain effects by means of formulae, and he represents forces by these formulae, then in the case where these forces are expressed by a negative sign, when pressure forces are made negative, they become suction forces.

Attention must be paid to the fact that in such a case one leaves the spatial realm entirely. Consideration of such formulae leads us into the realm of heat and light effects. Heat effects are only half included, for in the realm of heat we have both pressure and suction effects playing into each other. These facts, my dear friends, can be given only theoretically today, in this presentation in an auditorium. It must never be forgotten, however, that a large part of our technological achievement has arisen under the influence of materialistic concepts of the second half of the nineteenth century. Modern technology did not have ideas such as we are now presenting, and therefore such ideas cannot arise in it. If you consider how fruitful those one-sided concepts have been for technology, you can imagine how many technical consequences might flow from adding to modern technology — which only takes account of pressure forces — the possibility of making these suction forces fruitful also (and by these I mean not only spatially active suction, which is a manifestation of pressure, but suction forces *qualitatively opposite* to pressure forces). Of course, much now incorporated in the body of knowledge known as physics will have to be discarded to make room for these ideas. For instance, the usual concept of energy must be thrown out. This concept rests on the following very crude notions. When I have heat I can change it into work, as we saw in the experiment from the up-down movement of the piston resulting from the transformation of heat.[42] But we saw at the same time that the heat was only partly changed and that a portion remained, that only a portion was at our disposal to transform into what physics calls mechanical work, while the other portion could not be transformed in this way.[43] Other physicists — Mach, for example, who is well known in connection with modern developments in this field — have done fundamental thinking on the subject. Mach has thought along lines that show him to be a shrewd investigator but one who can only make use of his shrewdness under the influence of a purely materialistic view. Behind his concepts always stands the materialistic point of view. He shrewdly seeks to push forward and extend the concepts and ideas available to him. His noteworthy characteristic is that when he comes to the limit of the usual physical concepts, where doubts begin to arise, he writes the doubts down at once. This soon leads to despair, because he quickly arrives at the limit where doubts appear, but his way of expressing the matter is extremely interesting. Consider how things stand when a man who has the whole of physics at his command is obliged to state his views as Mach states them. He says, 'There is no sense in expressing as a work value a heat quantity that cannot be transformed into work.' (We have seen that there is such a residue.) 'Thus it appears that the energy principle, like other knowledge of substance, is valid only for a limited realm of facts. The existence of these limits is a matter about which we ordinarily are happy to deceive ourselves.'

Consider a physicist who, upon thinking over the phenomena lying before him, is obliged to say:

There is in fact heat that I cannot turn into work. But there is no meaning in simply thinking of this heat as potential energy, as work that is simply not visible. I can perhaps speak of the transformation of heat into work within a certain realm, but beyond this realm it is no longer valid. Generally it is said that every energy is transformable into another, but this can be accepted only by virtue of a certain habit of thinking about which we gladly deceive ourselves.

It is extremely interesting to pin physics down at the very point where doubts are expressed, doubts that arise necessarily from a straightforward consideration of the facts. When physicists are obliged to make such confessions, does this not clearly reveal the way in which physics itself falls short? For the energy principle is fundamentally nothing but a conjecture. One can no longer hold to the energy principle put forth as gospel by Helmholtz and his contemporaries. There are realms in which this energy principle can no longer be upheld.

Now let us ask the following: How can we attempt to symbolize what occurs in the realm of heat? When you bring together all these ideas that I have developed here, and through which I have tried to reach the nature of heat in a real sense, you come to a concept of heat in the following way.

Picture to yourselves [see Fig. 18] that here there is space (blue) filled with certain effects, pressure effects; here is the non-spatial (red) filled with suction effects.

If you picture this now, you come here to a realm that is different, that is always slipping in here and disappearing—we have projected out into space what can only be thought of as spatial/non-spatial, for the red portion must be thought of as non-spatial. You see the space here (blue and red) as a symbol of what is spatial/non-spatial. Think of something represented as 'extensive' and 'intensive',

blue

red

*Fig. 18*

through which materiality continually arises. As substance arises there enters in something from the other side that is immaterial, that slips into the substance and annihilates it; then we have a physical-spiritual vortex continually manifesting in such a way that what appears physically is annihilated by what appears as the spiritual. Therefore we have a vortex effect in which the physical comes into being that is then annihilated by the spiritual. We have a continuous interplay between the non-spatial and the spatial. We have a continual sucking up of what exists in space by the entity which is outside space.

You must think of what I am describing to you here as shaped similarly to a vortex. But in this vortex you should see simply a visible extension of that which is 'intensive' in its nature. In this way we approach, I might say figuratively, the nature of heat. We have yet to show how this nature of heat works so as to bring about such phenomena as conduction or the lowering of the melting-point of an alloy below the melting-point of its constituents, and what it really means that we should have heat effects at one end of the spectrum and chemical effects at the other.

We must seek the *deeds of heat* as Goethe sought out the *deeds of light*.[44] Then we must see how knowledge of the nature of heat is related to the application of mathematics and how it affects the imponderables of physics. In other words, how are real mathematical formulae to be constructed that can be applied to heat and optics.[45]

# 11. The Working of the Ethers in the Physical

*Are the laws of physics eternal? For all practical purposes, this is how they are viewed, although modern cosmology postulates that in the earliest moments of the 'Big Bang' space, time and the laws of physics were still being formed. In other words, it has proved impossible to apply the known laws of physics consistently and thereby explain the origin of the universe.*

*Rudolf Steiner reaches a similar conclusion but using an entirely different approach. By considering the scale of nature, with heat at the centre, he examines the relationship between the realms above heat (the 'ethers') and the states of matter below it. This leads him to conclude that all of the relationships now obtaining in nature were once different. For example, heat once had a dual nature, and there was once a close kinship between light and air. Such descriptions remind us of the spiritual-scientific descriptions Steiner gives of the great evolutionary processes of the earth, for example as described in* An Outline of Esoteric Science *— see under 'Further Reading', p. 237.*

The heat realm, the x, y and z realms, the gaseous, fluid and solid realms, and the U realm are to be arranged as we have outlined. Recollect that there is, if we remain purely in the realm of phenomena, a certain very loose, mutual relationship to be observed between heat effects and phenomena manifested in a gaseous mass.[46] We are able to observe that the gaseous body manifests in its material configuration what the heat is doing. The activity of heat can be found in its material expression in what the gas is doing. If we cultivate a vivid insight into what occurs in this interplay between gaseous matter and heat, between the heat effects and their material manifestation in the gaseous realm, we will also be able to get a real concept of the difference between the x realm and the realm of gases. We need only consider what we have seen countless times in life, that what we call light does not relate itself in the same way to gas as does heat. Gas does not follow changes in light by corresponding changes in its material configuration. When light spreads out, the gas does not do likewise, it does not show a greater elasticity, etc.

Therefore when light is playing through a gas, the relationship is different from the one existing between gas and the heat playing through it. Now, in the observations made previously, we said: fluidity stands between gas and solid, heat between the x realm and the gaseous realm. Also the solid realm gives a picture of the fluid realm, the fluid realm gives a picture of the gaseous realm, and the gaseous realm gives a picture of the heat realm. So likewise we can say that the x realm can be a picture of heat, while heat is itself pictured in the gaseous realm. In the gaseous realm we have, as it were, pictures of pictures of the x realm.

Consider now that these pictures of pictures are really

present when light passes through the air. Considering how the air relates itself and its various phenomena to light, one must say that we are not dealing with a direct imaging of one realm by the other; rather, the light actually has an independent status in the air. This independent relationship may be compared to the following. Suppose we want to paint a landscape and hang the picture on the wall of this room and then photograph the room. By changing something in the room I alter its whole appearance, and this alteration is evident in the photograph. If I were accustomed always to sit on this chair when giving a lecture, and some ill-disposed person removed it while I lectured without my noticing what he was doing, I would do what many have done under similar circumstances, namely, sit on the floor. The relation of things in the room undergoes real changes when I alter something in the room. But whether I hang the picture in one place or another, the relationship between the various figures painted in the picture does not change. What exists in the picture itself in the way of relationships is independent of alterations that go on in the room. In the same way, my experiments with light are independent of the air in the space in which they are carried out. Experiments with heat are, on the contrary, not independent of the space in which they are carried out, as you can easily convince yourselves; indeed, you are made aware of this by the whole room becoming warm. But my light experiments have an independent being; I can picture them in relief as it were,[47] so that when I experiment with x in an air-filled space, the same relationships exist as when I experiment with light. I can identify x with light.

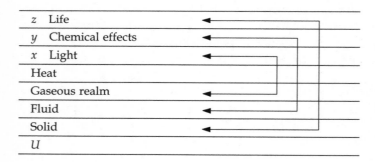

| z | Life |
| y | Chemical effects |
| x | Light |
| Heat | |
| Gaseous realm | |
| Fluid | |
| Solid | |
| U | |

A further extension of this train of thought leads to the identification of y with chemical effects and of z with life effects. However, as you see, there is a certain independence between the light realm and the gaseous realm. The same sort of relationships are found if we extend the train of thought — you can do it for yourselves. It would lead us too far to do it here today — for instance, if we look for the chemical effects in the fluid realm. In fact, in order to call forth chemical action, solutions are always necessary. In these solutions chemical action is related to the fluid as light is to air, to the gaseous state. We would then expect to find the z realm associated with the solid. If I indicate the three realms by z, y and x, with heat as the intermediate condition, I can then designate the gaseous realm with x', the fluid realm with y', and the solid realm with z'; I can represent the order like this:

z, y, x, heat, x', y', z', where

x in x' represents light in gas,
y in y' represents chemical effects in fluids, and
z in z' represents the z effects in solid bodies. Formerly we knew them only as forms.

Thus we get intermingling, as it were, but these are nothing but conceptual expressions of things that are very real in our lives:

x in x' is simply light-filled gas,
y in y' is fluidity in which chemical processes are going on, and
z in z' is life effects acting in solids.[48]

Just as we proceed beyond heat to find light, just as we proceed from light to find chemical effects, so beyond the chemical effects we must come to the life effects. Therefore we may say that z in z' represents life effects in solid bodies. But there is no such thing as life effects in solid bodies. We know that under earthly conditions a certain degree of fluidity is necessary for life. Under earthly conditions life effects are not present in the purely solid state. But these earthly conditions force us to establish the hypothesis that such a condition is not beyond the realm of possibility, for the order in which we have been able to think of these things necessarily leads to this.

We find solid bodies, we find fluid bodies, we find gas. The solids we find without life effects. Life effects in the earthly sphere we only discover unfolding themselves adjacent to solid bodies, in relation to them, etc. But in the earthly realm we do not find a direct coupling of life effects with what we call solids. We are led to this last member of the series z in z′, life in the solid realm, by analogy with y in y′ and x in x′. If I have a fluid body on the earth, it must have the same relation to chemical activity, although not so strong, as solid bodies do to life. Gases in the earthly realm must stand in the same relation to light as solids do to the living. Now, this leads us to recognize that in the earthly domain solids, fluids and gases in their further relations to light, chemical action and life phenomena represent something that has died away.

These thoughts cannot be made as obvious as people like to make most presentations of empirical facts. If you wish to make these facts really connect with reality, you must work them over within yourselves, and then, if you continue this sequence of thoughts, you will find that there is a kinship between the solid and the living, the fluid and the chemical, and the gaseous and light.

Heat stands independently of the other realms in a certain way. These relationships are not, however, directly expressed under earthly conditions. The relationships that can exist in the earthly domain point to something that was once there but is there no longer. Certain inner relationships in these things force us to introduce time concepts into the picture. When you look at a corpse you are forced to introduce concepts of time. The corpse is there but everything that makes possible the presence of the corpse that gives it its appearance, all this you must consider as the soul-spiritual element since the corpse has in itself no possibility of existence. The form of the human body would never arise without the soul-spiritual element. What the corpse presents to you forces you to say that what is there has been abandoned by something. This is no different from saying: the earthly solid has been abandoned by life, the earthly fluid by the emanations of the chemical effects, the earthly gaseous by the emanations of light effects. And just as we look back from the corpse to life, to the time when matter that is now the corpse was bound together with the soul-spiritual element, so we look from the solid bodies of the earth back to a former physical condition, when the solid was bound up with the living.

At that time the entire earth was not solid as we now understand the solid condition, just as little as was the corpse of today a corpse five days ago. Solids were not found in an independent state anywhere on the earth and only occurred bound to what was living; fluid existed only bound to chemical effects and gases only bound to light effects. In other words, all gas had an inner glittering, an inner illumination, an illumination that showed a wave-like phosphorescence and darkening as the gas rarefied or condensed. Fluids were not as they are today but were permeated by a continuous living chemical activity. And at the foundation of all was life, solidifying itself (as it solidifies now in the horn formation in cattle, for instance) and then passing back again into fluid or gas, etc. In brief, we are forced by physics itself to admit a previous period of time when realms now sundered on the earth existed in a simultaneous, interconnected way. The realms of the gaseous, the fluid and the solid are now found on the one hand, and on the other hand the realms of light, chemical effects and life. At that time they were within each other, not merely side by side, but actually within each other.

Heat had an intermediate position. It did not appear to share this kinship of the material and the more etheric natures. But since it occupied an intermediate position, it is clearly conceivable that it participates in both the material and the etheric. If we now call the upper realm the etheric[49] and the lower realm that of ponderable matter, we obviously have to consider the heat realm as the condition of equilibrium between them. Thus in heat we have found the condition of equilibrium between the etheric and the ponderable material realm. It is ether and matter at the same time and indicates by its dual nature what we actually find everywhere in heat, that is, a variation in level. This is an observation without which we cannot understand or arrive at anything in the realm of heat phenomena.

If you take up this line of thinking, you will come to something much more fundamental and important than the so-called second law of thermodynamics. For this second law really sunders a certain realm of phenomena from its proper connection; this realm is bound up with other phenomena and is essentially and profoundly modified by them.

If you make it clear to yourselves that the gaseous realm and the light realm were once one, that the fluid realm and the chemical effects were once one, etc., then you will also be led to think of the two polar opposite portions of the heat realm, namely, ether and ponderable matter, as originally united. That is to say, you must conceive of heat in former ages as completely different from heat as you think of it now. Then you will come to say to yourselves that the things we designate as physical phenomena today, the expression of which is in their physical entity, are limited in their meaning by time. Physics is not eternal. For many completely different types of reality physics has absolutely no validity. For of course the reality in which gas was once directly illumined within is an entirely different reality from that in which gas and light are relatively independent in relation to one another.

Thus we come to look back on a time when another type of physics was valid; and looking into the future, there will be a time when a still different type of physics will come into being. Our modern physics can only conform with the phenomena of the present time, with what is in our immediate environment. In order to avoid paradoxes, and not only these but absurdities, physics itself must be freed of the tendency to study physical phenomena from our earthly perspective, build hypotheses based on them, and then apply these hypotheses to the whole universe.

We apply our earthly hypotheses to the universe and forget that what we know as physical is limited by time to the earthly domain. That it is limited in space we have already seen, for the moment we move out to the sphere where gravity ceases and everything streams outwards, at that moment our entire physical view of the world ceases to apply.

We have to say, therefore, that our earth is not simply spatially limited but that it is spatially limited in its physical qualities. It is nonsensical to suppose that beyond the zero- or null-sphere [*see Fig. 19*] something will be found to which the same physical laws must apply. It is just as nonsensical to apply the same physical laws to former ages and to derive the nature of earthly evolution from what is going on at a particular time. The madness of the Kant-Laplace theory consists of the belief that it is possible to abstract something from contemporary physical phenomena and extend it arbitrarily backwards in time. Modern astrophysics also shows the same madness in the belief that what can be abstracted from earthly

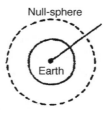

*Fig. 19*

physical conditions can be applied to the constitution of the sun, for example, and that we can speak about the sun on the basis of earthly physical laws.

But something tremendously important presents itself to us if we take this overview of the phenomena that we have gained and then bring together certain series of phenomena. Your attention has been drawn to the fact that physicists have come to a view neatly expressed by Eduard von Hartmann. The second law of thermodynamics states that whenever heat is changed into mechanical work some heat remains unchanged, and thus, finally, all energy must change into heat and the earth comes to a heat death. This view has been expressed by Eduard von Hartmann as follows: 'The world process has the tendency to run down.' Now suppose we assume that such a running down of the world process does take place in the direction indicated, what happens then?

When we conduct experiments to illustrate the second law of thermodynamics, we see that heat appears; we see mechanical energy used up and heat appearing. What we see appearing undergoes a further change. For we can show likewise, when we produce light from heat, that not all of the heat reappears as light, since heat simply reverses the mechanical process as it is understood in the sense of the second thermodynamic law. It is similar with the relationship between light phenomena and chemical phenomena.

This has led us, however, to say that we have to imagine the whole cosmic spectrum as bent around into a circle.

Thus if it were really true, as examination of a certain series of phenomena indicates, that the entropy of the universe is striving to reach its maximum and that the world process is running down, provision is also made for it to run back. It runs down here on this side but then runs back from here [*indicating figure*] on the other side, for we have to think of it as a circle. Thus even if heat-death[50] actually occurs on one side, on the other side there occurs something to re-establish the equilibrium and that opposes the world's death with universal creation. This follows from a sober observation of the phenomena.

This can be verified in physics if it no longer observes the world process in the same linear way that we usually observe the sun's spectrum—tracing red into infinity in the past on one side, and tracing blue into infinity in the future on the other side.

Instead we must symbolize the world process in a circular way, and only then can we draw near to the world process. Once we have symbolized the world process as a circle, however, then we can include in it what lies in our various realms.[51]

## The ethers in the human organism

How are thoughts produced? Through the fact that within the fluid organism something asserts itself in a particular metamorphosis, namely, what we know in the external world as tone. Tone is in reality something that leads our ordinary mode of observation very much astray. As earthly human beings we perceive tone as being borne to us on the air. But in point of fact the air is only the transmitter of the tone, which actually weaves in the air. And anyone who assumes that tone in its essence is merely a matter of air vibrations is like a person who says: Man has only his physical organism, and there is no soul in it. If the air vibrations are thought to constitute the essence of tone, whereas they are in truth merely its external expression, this is the same as seeing only the physical organism with no soul in it. The tone which lives in the air is essentially an etheric reality. And the tone we hear by way of the air arises through the fact that the air is permeated by the tone ether, which is the same as the chemical ether. In permeating the air, this chemical ether imparts what lives within it to the air, and we become aware of what we call the tone.

This tone ether or chemical ether is essentially active in our fluid organism. We can therefore make the following distinction: in our fluid organism lives our own etheric body, but in addition there penetrates from every direction into the fluid organism the tone ether which underlies tone. Please distinguish carefully here. We have within us our etheric body; it works and is active by giving rise to thoughts in our fluid organism. But what may be called the chemical ether continually streams in and out of our fluid organism. Thus we have an etheric organism complete in itself, consisting of chemical ether, warmth ether, light ether, life ether, and in addition we find in it, in a very special sense, the chemical ether which streams in and out by way of the fluid organism.

The astral body, which comes to expression in feeling, operates through the air organism. But still another kind of ether permeating the air is connected especially with the air organism. This is the light ether. Earlier conceptions of the world always emphasized this affinity of the outspreading physical air with the light ether which permeates it. This light ether, which is borne, as it were, by the air and is related to the air even more intimately than tone, also penetrates into our air organism and underlies what there passes into and out of it. Thus we have our astral body, which is the bearer of feeling, is especially active in the air organism, and is in constant contact there with the light ether.

And now we come to the ego. This human ego, which by way of the will is active in the warmth organism, is again connected with the outer warmth, with the in-streaming and out-streaming warmth ether.

Thus we obtain the following relations:

Ego — will — warmth organism — warmth ether
Astral body — feeling — air organism — light ether
Etheric body — thinking — fluid organism — chemical ether

# 12. Sub-nature

*One of the most remarkable insights in Steiner's view of nature is the concept of a part of nature which, through the working of certain forces, is separated from the usual cosmic dimension. Sub-natural forces include magnetism and electricity, which are of course the very forces used in modern technology. Steiner explains how sub-nature arises, and warns us that we must learn how to use it in a way worthy of the true goals of mankind, if it is not to lead us downwards into barbarism and cultural or spiritual decline.*

*There is little information on sub-nature. The first extract here is from a lecture on the role of Christ in the evolution of the earth. Steiner gives little more than an indication of how these forces are possible as the earth gradually becomes spiritualized. Then, after the lecture, he provides more information in response to questions from the audience on the nature of chemical effects, and electricity. It is easy to misunderstand Steiner's words – in recent times critics of anthroposophy have interpreted some phrases of this lecture as meaning that spiritual science wishes to see the earth destroyed so that we may be free! One might just as well accuse meteorologists of wanting to see the earth destroyed by flood and hurricane.*

*The second extract comes from a written, as opposed to verbal, source, and is therefore worded more circumspectly. It also gives a practical method for coping with a potentially destructive technology: working to achieve a corresponding increase in spiritual development.*

If we investigate early post-Atlantean times, we find that human beings built their dwelling-places by methods very different from those used nowadays. In those days they made use of all kinds of growing things. Even when building palaces they summoned nature to their aid by utilizing plants interlaced with branches of trees and so on, whereas today people must build with broken fragments. All the culture of the external world is contrived with the aid of products of fragmentation.[52] And in the course of the coming years you will realize even more clearly how much in our civilized life is the outcome of destruction.

Light itself is being destroyed in this post-Atlantean age of the earth's existence, which until the time of Atlantis was a progressive, life-creating process. Since then it has been a process of decay. What is light? Light decays and the decaying light is electricity. What we know as electricity is light that is being destroyed in matter.[53] And the chemical force that undergoes a transformation in the process of earth evolution is magnetism. Yet a third force will become active; and if electricity seems to work wonders today, this third force will affect civilization in a still more miraculous way. The more of this force we employ, the faster will the earth tend to become a corpse and its spiritual part prepare for the Jupiter embodiment.[54]

Forces have to be applied to destructive purpose in order that man may become free of the earth and that the earth's body may fall away. As long as the earth was involved in progressive, life-creating evolution, no such destruction took place, for the great achievements of electricity can only serve a decaying earth. Strange as this sounds, it must gradually become known.

## What is the relation of chemical forces and substances to the spiritual world?

There are in the world a number of substances which can combine with or separate from each other. What we call chemical action is projected into the physical world from the world of Devachan—the realm of the harmony of the spheres. In the combination of two substances according to their atomic weights, we have a reflection of two tones of the harmony of the spheres. The chemical affinity between two substances in the physical world is like a reflection from the realm of the harmony of the spheres. The numerical ratios in chemistry are an expression of the numerical ratios of the harmony of the spheres, which has become dumb and silent owing to the densification of matter. If man were able to etherealize material substance and to perceive in the atomic numbers the inner formative principle working there, he would hear the harmony of the spheres.

We have the physical world, the astral world, the Lower

Devachan and the Higher Devachan.[55] If something is thrust down lower even than the physical world, it comes into the sub-physical world, the lower astral world, the lower or evil Lower Devachan and the lower or evil Higher Devachan. The evil astral world is the province of Lucifer, the evil Lower Devachan the province of Ahriman, and the evil Higher Devachan the province of the Asuras. When chemical action is driven down beneath the physical plane — into the evil devachanic world — magnetism arises. When light is thrust down into the sub-material — that is to say a stage lower than the material world — electricity arises. If what lives in the harmony of the spheres is thrust down farther still, into the province of the Asuras, an even more terrible force, which it will not be possible to keep hidden very much longer, is generated.[56] It can only be hoped that when this force comes to be known — a force we must conceive as being far, far stronger than the most violent electrical discharge — it can only be hoped that before some inventor gives this force into the hands of humankind, people will no longer have any immorality left in them.

## What is electricity?

Electricity is light in the sub-material state. Light is there compressed to the utmost degree. An inward quality too must be ascribed to light; light is itself at every point in space. Warmth will expand in the three dimensions of space. In light there is a fourth; it is of fourfold nature and has the quality of inwardness as a fourth dimension.

Reflected as sub-physical world:

Astral world — the province of Lucifer
Lower Devachan — the province of Ahriman
Higher Devachan — the province of the Asuras

Life ether
Chemical ether
Light ether
Sub-physical astral world — electricity
Sub-physical Lower Devachan — magnetism
Sub-physical Higher Devachan — terrible forces of destruction

## From nature to sub-nature

The Age of Philosophy is often said to have been super-seded, about the middle of the nineteenth century, by the rising Age of Natural Science. And it is said that the Age of Natural Science still continues in our day, although many people are at pains to emphasize at the same time that we have found our way once more to certain philosophic tendencies.

All this is true of the paths of knowledge which the modern age has taken, but not of its paths of life, its ways of living. With his conceptions and ideas, man still lives in nature, even if he carries the mechanical habit of thought into his theories about it. But with his life of will he lives in the mechanical processes of technical science and industry to so far-reaching an extent that it has long imbued this Age of Science with an entirely new quality.

To understand human life we must first consider it from two distinct aspects. From his former lives on earth man brings with him the faculty to conceive the cosmic — the cosmic that works inwards from the earth's encircling spheres, and that which works within the earth domain itself. Through his senses he perceives the cosmic that is at work upon the earth; through his thinking organization he conceives and thinks the cosmic influences that work downwards to the earth from the encircling spheres.

Thus man lives, through his physical body in perception, through his etheric body in thought.

What takes place in his astral body and his ego holds sway in the more hidden regions of the soul. It holds sway, for example, in his destiny. We must, however, look for it, to begin with, not in the complicated relationships of destiny, but in the simple and elementary processes of life.

Man connects himself with certain earthly forces, in that he gives his body its right orientation within them. He learns to stand and walk upright; he learns to place himself with arms and hands into the equilibrium of earthly forces.

Now these are not forces working inwards from the cosmos. They are forces of a purely earthly nature.

In reality, nothing that man experiences is an abstraction. He only fails to perceive where experiences come to him from; and thus he turns ideas about realities into abstractions. He speaks of the laws of mechanics. He thinks he has abstracted them from the connections and relationships of nature. But this is not the case. All that man experiences in his soul by way of purely mechanical laws has been discovered inwardly through his orientation and relationship to the earthly world (in standing, walking, etc.).

The mechanical is thus characterized as that which is of a purely earthly nature. For the laws and processes of nature as they hold sway in colour, sound, etc., have entered into the earthly realm from the cosmos. It is only within the earthly realm that they too become imbued with the mechanical element, just as is the case with man himself, who does not confront the mechanical in his conscious experience until he comes within the earthly realm.

By far the greater part of what works in modern civilization through technical science and industry — with which the life of man is so intensely interwoven — is not nature at all, but sub-nature. It is a world that emancipates itself from nature, and does so in a downward direction.

Look how the oriental, when he strives towards the spirit, seeks to get out of the conditions of equilibrium whose origin is merely in the earthly realm. He assumes an attitude of meditation which brings him again into a purely cosmic balance. In this attitude the earth no longer influences the inner orientation of his body. (I am not recommending this for imitation; it is mentioned merely to make our present subject clear. Anyone familiar with my writings will know how different is the Eastern from the Western spiritual life.)

Man needed this relation to the purely earthly for the unfolding of his spiritual soul.[57] Thus in the most recent times there has arisen a strong tendency to realize in all things, and even in the life of action, this element into which man must enter for his evolution. Entering the purely earthly element, he comes up against the ahrimanic realm. With his own being he must now acquire a right relation to the ahrimanic.

But in the age of technical science hitherto, the possibility of finding a true relationship to ahrimanic civilization has escaped man. He must find the strength, the inner force of knowledge, in order not to be overcome by Ahriman in this technical civilization. He must understand sub-nature for what it really is. This he can only do if he rises, in spiritual knowledge, at least as far into the realm of super-nature, beyond the earthly, as he has descended, in technical sciences, into sub-nature. The age requires a knowledge that transcends nature, because in its inner life it must come to grips with a life-content that has sunk far beneath nature — a life-content whose influence is perilous. Needless to say, there can be no question here of advocating a return to earlier states of civilization. The point is that man needs to find the way to bring the conditions of modern civilization into their true relationship — to himself and to the cosmos.

There are very few as yet who even feel the greatness of these spiritual tasks approaching man. Electricity, for instance, celebrated since its discovery as the very soul of nature's existence, must be recognized in its true character — in its peculiar power of leading down from nature to sub-nature. Only man himself must beware lest he slide downwards with it.

In the age when there was not yet a technical industry independent of true nature, man found the spirit within his view of nature. But technical processes, emancipating themselves from nature, caused him to stare more and more fixedly at the mechanical-material, which now became for him the really scientific realm. In this mechanical-material domain, all divine-spiritual being connected with the origin of human evolution is completely absent. The purely ahrimanic dominates this sphere.

In the science of the spirit we can now strive to create another sphere in which there is no ahrimanic element. It is by receiving in knowledge this spirituality to which the ahrimanic powers have no access that man is strengthened to confront Ahriman within the world.

# 13. What are Atoms?

*Rudolf Steiner made many comments about the atomic theory of his day, some of them quite disparaging — indeed, it would be easy to form the impression that he did not accept the existence of atoms! However, it must be noted that although atomistic theories of matter were largely the accepted currency in his lifetime (and totally undisputed today), there were still several eminent authorities who held out against them, such as the Nobel Prizewinner Friedrich Ostwald. What Steiner really objected to was the thinking of his day, which postulated atoms without good reason (see Chapter 4). This is evident in his short essay of 1890, 'Atomism and its Refutation', where he shows the inconsistency of attributing the qualities of the sense-world (such as sound and colour) to the motions of atoms which have none of these qualities. Later, in 1904, he described the atom as 'congealed electricity',[58] a description that would be possible for a modern chemist to accept. He also describes atoms as being a focus of forces. Not so easy to understand is the view revealed by imaginative cognition, that atoms are empty bubbles, i.e., they contain no spiritual reality, but are filled with spiritless substance.*

*A wider prospect is revealed when we see 'atomism' as a stream of thought which analyses phenomena into ever-smaller components, thereby missing the essential connections between the parts needed to understand organisms. Atomism in this sense can only grasp inorganic nature. In contrast, seeing nature as a continuous, connected whole — 'continuism' — can grasp the organic, living aspect.*

## What did Democritus mean by 'atoms'?

There is a great difference between modern-day atomists and Democritus.[59] His utterances were based on the awareness of the contrast between man and nature, soul and body [*see Chapter 1*]. His atoms were complexes of force and as such were contrasted with space, something a modern atomist cannot do in that manner. How could the modern atomist say what Democritus said: 'Existence is not more than nothingness, fullness is not more than emptiness?' It implies that Democritus assumed empty space to possess an affinity with atom-filled space. This has meaning only within a consciousness that as yet has no idea of the modern concept of body. Therefore, it cannot speak of the atoms of a body, but only of centres of force, which, in that case, have an inner relationship to what surrounds man externally. Today's atomist cannot equate emptiness with fullness. If Democritus had viewed emptiness the way we do today, he could not have equated it with a state of being. He could do so because in this emptiness he sought the soul that was the bearer of the Logos. And though he conceived this Logos in a form of necessity, it was the Greek form of necessity, not our modern physical necessity. If we are to comprehend what goes on today, we must be able to look in the right way into the nuances of ideas and feelings of former times.

## Atomism versus continuism

August Weismann, a biologist of the nineteenth century, conceived the thought that in any living organism the *interplay* of the organs (in lower organisms, the interaction of the parts) must be regarded as the essential thing. This leads to comprehension of how the organism lives. But in examining the organism itself, in understanding it through the interrelationship of its parts, we find no equivalent for the fact that the organism must die. If one only observes the organism, so Weismann said, one finds nothing that will explain death. In the living organism, there is absolutely nothing that leads to the idea that the organism must die. For Weismann, the only thing that demonstrates that an organism must die is the existence of a corpse. This means that the concept of death is not gained from the living organism. No feature, no characteristic, found in it indicates that dying is a part of the organism. It is only when the event occurs, when we find a corpse in the place of the living organism, that we know the organism possesses the capacity to die.

But, says Weismann, there is a class of organisms where corpses are never found. These are the unicellular organisms. They only divide themselves so there are no corpses. The propagation of such beings looks like this [*see Fig. 20*]:

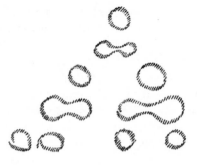

*Fig. 20*

One divides into two; each of these divide into two again, and so on. There is never a corpse. Weismann therefore concludes that unicellular beings are immortal. This is the immortality of unicellular beings that was famous in nineteenth-century biology. Why were these organisms considered immortal? Because they never produce any corpses, and because we cannot entertain the concept of death in the organic realm as long as there are no corpses. Where there is no corpse, there is no room for the concept of death. Hence, living beings that produce no corpses are immortal.

This example shows how far man has removed himself in modern times from any connection between the world and his thinking, his inner experiences. His concept of an organism is no longer such that the fact of its death can be perceived from it. This can only be deduced from the existence of something like a corpse. Certainly, if a living organism is only viewed from without, if one cannot experience what is in it, then indeed one cannot find death in the organism and an external sign is necessary. But this only proves that in his thinking man feels himself separated from the things around him.

From the uncertainty that has entered all thinking about the corporeal world, from this divorce between our thoughts and our experience, let us turn back to the time when self-experience still existed. Not only did the inwardly experienced concept exist alongside the externally derived concept of a triangle, square or pentagram, but there were also inwardly experienced concepts of blossoming and fading, of birth and death. This inner experience of birth and death had its gradations. When a child was seen to grow more and more animated, when its face began to express its soul, when one really entered into this growing process of the child, this could be seen as a continuation of the process of birth, albeit a less pronounced and intensive one. There were degrees in the experience of birth. When a person began to show wrinkles and grey hair and grew feeble, this was seen as a first mild degree of dying. Death itself was only the sum total of many less pronounced death experiences, if I may use such a paradox. The concepts of blossoming and decaying, of being born and dying, were inwardly alive.[60]

These concepts were experienced in communion with the corporeal world. No line was drawn between man's self-experience and events in nature. Without a coastline, as it were, the inner land of man merged into the ocean of the universe. Owing to this form of experience, man lived into the world itself. Therefore the thinkers of earlier ages, whose ideas no longer receive proper attention from science, had to form quite different ideas concerning something like what Weismann called the 'immortality of unicellular beings'.

What sort of concept would an ancient thinker have formed had he had a microscope and known something about the division of unicellular organisms? He would have said: First I have the unicellular being; it divides itself into two. Somewhat imprecisely, he might have said: It atomizes itself, it divides itself; for a certain length of time, the two parts are indivisible, then they divide again. As soon as division or atomization begins, death enters in. He would not have derived death from the corpse but from atomization, from the division into parts. His train of thought would have been somewhat as follows.

A being that is capable of life, that is in the process of growth, is not atomized; and when the tendency to atomization appears, the being dies. In the case of unicellular beings, he would simply have thought that the two organisms cast off by the first unicellular being were for the moment dead, but would be, so to speak, revived immediately, and so forth. With atomization, with the process of splitting, he would have linked the thought of death. If he had known about unicellular beings and had seen one split into two, he would not have thought that two new ones had come into being. On the contrary, he would have said that out of the living monad, two atoms have originated. Further, he would have said that wherever there is life, wherever one observes life, one is not dealing with atoms. But if they are found in a living being, then a proportionate part of the being is dead. Where atoms are found, there is death, there is something inorganic. This is how matters would have been judged in a former age based on living inner knowledge of the world.

All this is not clearly described in our histories of philosophy, although the discerning reader can have little doubt of it. The reason is that the thought-forms of this older philosophy are totally unlike today's thinking. Therefore anyone writing history nowadays is apt to put his own modern concepts into the minds of earlier thinkers. But this is unallowable even with someone as recent as Spinoza. In his book on what he justifiably calls ethics, Spinoza follows a mathematical method but it is not mathematics in the modern sense. He expounds his philosophy in a mathematical style, joining idea to idea as a mathematician would. He still retains something of the former qualitative experience of quantitative mathematical concepts. Hence, even in contemplating the qualitative aspect of man's inner life, we can say that his style is mathematical. Today with our current concepts, it would be sheer nonsense to apply a mathematical style to psychology, let alone ethics.

If we want to understand modern thinking, we must continually recall this uncertainty, contrasting it to the certainty that existed in the past but is no longer suited to our modern outlook. In the present phase of scientific thinking, we have come to the point where this uncertainty is not only recognized but theoretical justifications have been offered for it. An example is a lecture given by the French thinker Henri Poincaré in 1912 on current ideas relating to matter. He speaks of the existing controversy or debate concerning the nature of matter, whether it should be thought of as being continuous or discrete. In other words, whether one should conceive of matter as substantial essence that fills space and is nowhere really differentiated in itself, or whether substance, matter, is to be thought of as atomistic, signifying more or less empty space containing within it minute particles that by virtue of their particular interconnections form into atoms, molecules, and so forth.

Aside from what I might call a few decorative embellishments intended to justify scientific uncertainty, Poincaré's lecture comes down to this. Research and science pass through various periods. In one epoch, phenomena appear that cause the thinker to picture matter in a continuous form, making it convenient to conceive of matter this way and to focus on what shows up as continuity in the sense data. In a different period the findings point more towards the concept of matter being diffused into atoms, which are pictured as being fused together again, i.e., matter is not continuous but discrete and atomistic. Poincaré is of the opinion that always, depending on the direction that research findings take, there will be periods when thinking favours either continuity or atomism. He even speaks of an oscillation between the two in the course of scientific development. It will always be like this, he says, because the human mind has a tendency to formulate theories concerning natural phenomena in the most convenient way possible. If continuity prevails for a time, we get tired of it. (These are not Poincaré's exact words, but they are close to what he really intends.) Almost unconsciously, as it were, the human mind then comes upon other scientific findings and begins to think atomistically. It is like breathing, where exhalation follows inhalation. Thus there is a constant oscillation between continuity and atomism. This merely results from a need of the human mind and according to Poincaré says nothing about the things themselves. Whether we adopt continuity or atomism determines nothing about things themselves. It is only our attempt to come to terms with the external corporeal world.

It is hardly surprising that uncertainty should result from an age which no longer finds self-experience in harmony with what goes on in the world but regards it only as something occurring inside man. If you no longer experience a living connection with the world, you cannot experience continuity or atomism. You can only impose your preconceived notions of continuity or atomism on natural phenomena. This gradually leads to the suspicion that we formulate our theories according to our changing needs. Just as we must breath in and out, so we must, supposedly, think first 'continuistically' for a while, then 'atomistically' for a while. If we always thought in the same way, we would not be able to catch a breath of mental air. Thus our fatal uncertainty is confirmed and justified. Theories begin to look like arbitrary whims. We no longer live in any real connection with the world. We merely think of various ways in which we might live with the world, depending on our own subjective needs.

What would the old way of thought have said in such a case? It would have said: In an age when the leading thinkers see things in a continuous way, they are thinking mainly of life. In one in which they think atomistically, they are thinking primarily of death, of inorganic nature, and they view even the organic in inorganic terms.

This is no longer unjustified arbitrariness. This rests on an objective relationship to things. Naturally, I can take turns in dealing with the animate and the inanimate. I can say that the very nature of the animate requires that I conceive of it in a continuous way, whereas the nature of the inanimate requires that I think of it atomistically. But I cannot say that this is only due to the arbitrary nature of the human mind. On the contrary, it corresponds to an objective way of relating oneself to the world. For such perception, the subjective aspect is really disregarded, because one recognizes the animate in nature in continuous form and the inanimate in discrete form. And if one really has to oscillate between the two forms of thought, this can be turned in an objective direction by saying that one approach is suited to the living and the other is suited to the dead. But there is no justification for making everything subjective as Poincaré does. Nor is the subjective approach valid for the way of perception that belonged to earlier times.

The gist of this is that in the phase of scientific thinking immediately preceding our own, there was a turn away from the animate to the inanimate, i.e., from continuity to atomism. This was entirely justified, if rightly understood. But, if we hope to objectively and truly find ourselves in the world, we must find a way out of the dead world of atomism, no matter how impressive it is as a theory. We must get back to our own nature and comprehend ourselves as living beings. Up to now, scientific development has tended in the direction of the inanimate, the atomistic. When, in the first part of the nineteenth century, this whole dreadful cell theory of Schleiden and Schwann made its appearance, it did not lead to continuity but to atomism. What is more, the scientific world scarcely admitted this, nor has it to this day realized that it should admit it since atomism harmonizes with the whole scientific methodology. We were not aware that by conceiving the organism as divided up into cells, we actually atomized it in our minds, which in fact means killing it. The truth of the matter is that any real idea of organisms has been lost to the atomistic approach.

## Spiritual perception of atoms

Let us hold on to the fact that there are those whose speculations are mainly concerned with matter; they imagine that the world consists of atoms. How does this view compare with what spiritual science has to say? Certainly natural physical phenomena do lead us back to atoms, but what are these atoms? They reveal what they are at the moment the very first stage of spiritual perception has been attained. At the stage of imaginative perception atoms reveal what they truly are. I have spoken about this in various connections many years ago in public lectures. Those who speculate on matter come to the conclusion that space is empty and atoms whirl around in this empty space. Atoms are supposed to be the most solid entities in existence. That is simply not the case; the whole issue is based on illusion. To imaginative cognition atoms are revealed as bubbles, and reality is where the empty space is supposed to be. Atoms are blown up bubbles. In other words, in contrast to what surrounds them they are nothing. You know that where bubbles are seen in soda-water there is no water. Atoms are bubbles in that sense — where they are the space is hollow, nothing is there. And yet it is possible to push against this; impact occurs precisely because, in pushing against hollowness, an effect is produced. How can nothing produce an effect? Take the case of the space, practically empty of air, within an air-pump. There you see how air streams into nothingness. A wrong interpretation might imagine the empty space in the bulb of the air pump as containing a substance that forced in the air. That is exactly the illusion prevailing in regard to the atom. The opposite is true: atoms are empty, yet again not empty. There is after all something within these bubbles. And what is it? This is also something about which I have already spoken. What exists within the atom bubbles is ahrimanic substance. Ahriman is there. The whole system of atoms consists of ahrimanic substantiality. As you see this is a considerable metamorphosis of the ideas entertained by those who theorize about matter. Where in space they see something material we see the presence of Ahriman.

# 14. Natural Science and Spiritual Science

*Nowadays the individual and indeed the whole of society is shaped by concepts that arose from the scientific experiments of the last few centuries. The old instinctive and religious social forms have been largely replaced by more conscious concepts, some of which have proved socially disastrous (such as the now largely discredited doctrines of Marxism). Today, one sees 'competition' as a driving force, which can be traced back to Darwinian theories of the survival of the fittest. Rudolf Steiner did not wish to discard the results of scientific research – he saw it as a necessary stage in the evolution of consciousness. He nonetheless pointed to the need to transcend crude materialism and develop a truer, life-imbued science which can provide socially fruitful concepts once again. A right understanding of scientific methodology can lead us to higher forms of cognition, such as Imagination and Inspiration, which will become increasingly necessary if mankind is to solve the moral and ethical dilemmas resulting from new technologies which are able to manipulate life itself. As fruitful starting-points, Steiner speaks of the right understanding of mathematics, especially projective geometry, and the phenomenological approach pioneered by Goethe.*

## Scientific concepts in social life

Those of you who are acquainted with my books will not have failed to observe that I am ready to do full justice to and in no way deny or criticize unfavourably the discoveries of modern times through scientific methods of research. I fully recognize what has been done for the progress of humanity by the Copernican world-view, by the science of Galileo, the widening of the horizon of mankind by Giordano Bruno, and much besides. But side by side with modern technical science, with modern capitalism, a gradual change has come about in the old world-view. The new conception of the world has taken on a decidedly intellectual and above all a scientific character. It is true that some people find it hard to look facts straight in the face, but we need only recall the fact that the scientific world-conception which we now regard with pride has gradually developed, as we can show, out of old religious, artistic, aesthetic, moral conceptions of the world. These views possessed a certain impelling force applicable to life. One truth, especially, was peculiar to them all. They led man to the consciousness of the spirituality of his own nature. However we may regard those old views, we must agree that they spoke to man of the spirit, so that he felt within himself the living spiritual being as a part of the cosmic spiritual being pulsating throughout the world, weaving the web of the universe. In the place of this old conception, with its impelling social force, giving an impulse to life, another appeared, new and more scientific in its orientation. This new conception was concerned with more or less abstract laws of nature, and facts of the senses, outside man himself, abstract ideas and facts. Without detracting in the smallest degree from the value of natural science, we may ask: What does it bestow on humanity, especially what does it bestow on man in order to help him solve the riddle of his own existence? Natural science tells us much about the interdependence of the phenomena of nature; it reveals much regarding the physical constitution of the human being. But when it attempts to tell us anything about man's innermost being, science overreaches itself. It can give no answer to this question, and it shows ignorance of itself when it even attempts to answer it.

I do not by any means wish to assert that the common consciousness of humanity already has its source in the teachings of modern science. But it is profoundly true that the scientific mode of thought itself proceeds from a certain definite attitude of the modern human soul. He who can penetrate below the surface of life knows that, since the middle of the fifteenth century, something in the attitude of the human soul has changed, when we compare it with former times, and is still changing more and more, and he also knows that the conception of the world which we find typically expressed in scientific thought has been diffused increasingly over the whole human race, first over the cities, then all over the land. It is, therefore, no mere achievement of theoretical natural science of which we are speaking, but an inner attitude of the soul which has gradually taken possession of humanity as a whole since the dawn of modern times. It is a significant coincidence that this scientific world-conception made its appearance at the same time as capitalism and modern technical culture. Human beings were called away from their old handiwork and placed at a machine, crowded together in a factory. The machine at which they stand, the factory in which they are crowded together with their fellows, these, governed only by mechanical laws, have nothing to give us that has any direct relationship to ourselves as human beings. Out of the old handicrafts something flowed to people which gave answer to their queries about human worth and human dignity. The dead machine gives no answer. Modern industrialism is like a mechanical network spun about the human being, in the midst of which he stands; it has nothing to give him in which he can joyfully share, as did the work of the old handicrafts.

## Science and the individual

When people today turn to natural science in order to reach a satisfying answer to the question 'Where, as human being, is my place in the world?', then at best the natural-scientific world view will tell them how their physical bodies relate to world evolution as a whole. Today it is known, at least up to a point, where man's physical body belongs in the evolutionary process. But the natural-scientific world-view has absolutely nothing to say about how man's soul, let alone spirit, fits into world evolution. Compare for a moment the evolutionary process, as described by spiritual science, with that described by natural science. The natural-scientific theory of evolution leads to the animal kingdom (how this is arrived at is a separate issue). Spiritual science leads us back through the different phases of Earth evolution: through Ancient Moon evolution, Ancient Sun evolution to Ancient Saturn evolution.[61] It shows us that what lives within us as soul and spirit were germinally present already within Ancient Saturn evolution. Nothing physically was then present, except conditions of warmth. We find how we are related to this primordial warmth, pervaded through and through by the individual beings of the hierarchies who are still about us. We are placed within a cosmos filled with soul and spirit. That is the great difference.

Spiritual science shows our soul and spirit to be part and parcel of a universal all, which it can describe in detail. Thus spiritual science alone can give the human soul that without which it feels annihilated. The dissatisfaction and insecurity felt by modern man reflect modern thinking. This thinking disregards the soul and declares that only the human body exists within the cosmic all. Another aspect is that the soul feels it has nothing to relate to, and that prevents it from finding inner strength. To reach inner strength of soul one must have attained concepts and ideas which depict the cosmic all as containing man as a being of soul and spirit; just as natural science depicts physical man as part of the physical evolution of the universe.

## Natural science as a foundation for spiritual science

We must begin by acquiring the discipline that modern science can teach us. We must school ourselves in this way and then, taking the strict methodology, the scientific discipline we have learned from modern natural science, transcend it, so that we use the same exacting approach to rise into higher regions, thereby extending this methodology to the investigation of entirely different realms as well. For this reason I believe—and I want this to be expressly stated—that nobody can attain true knowledge of the spirit who has not acquired scientific discipline, who has not learned to investigate and think in laboratories according to the modern scientific method. Those who pursue spiritual science have less cause to undervalue modern science than anyone. On the contrary, they know how to value it at its full worth. And many people—if I may here insert a personal remark—were extremely upset with me when, before publishing anything about spiritual science as such, I wrote a great deal about the problems of natural science in a way that appeared necessary to me. So you see it is necessary on the one hand for us to cultivate a scientific habit of mind, so that this can accompany us when we cross the frontiers of natural science. In addition, it is the quality of this scientific method and its results that we must take very seriously indeed.

## Mathematics: origins and future possibilities

We bear within ourselves three inner senses: the sense of life, the sense of movement, and the sense of balance. They are especially active in childhood up to the change of teeth. Around this time of the change of teeth their activity begins to wane, but observe, to take but one example, the sense of balance — observe how at birth the child has as yet nothing enabling it to attain the position of balance it needs in later life. Consider how the child gradually gains control of itself, how it learns at first to crawl on all fours, how it gradually achieves through its sense of balance the ability to stand and to walk, how it finally is able to maintain its own balance.

If one considers the entire process of development from conception to the change of teeth, one sees in it the powerful activity of these three inner senses. And if one can attain a certain insight into what is happening there, one sees that there is at work in the sense of balance and the sense of movement nothing other than a living 'mathematical activity'. In order for it to come to life, the sense of life is there to vitalize it. We thus see a kind of latent realm of mathematics active within man. This activity does not entirely cease at the change of teeth, but it does become at that time considerably less pronounced for the remainder of life. That which is inwardly active in the sense of balance, the sense of movement and the sense of life becomes free. This latent mathematics becomes free, just as latent heat can become liberated heat. And we see how what initially was woven through the organism as an element of soul becomes free. We see how this mathematics emerges as abstraction from a condition in which it was originally a concrete force shaping the human organism. And because as human beings we are suspended in the web of existence according to temporal and spatial relationships, we take this mathematics that has become free out into the world and seek to comprehend the external world by means of something that worked within us up until the change of teeth. You see, it is not a denial but rather an extension of natural science that results when one considers rightly what ought to live within spiritual science as attitude and will.

We thus carry what originates within ourselves beyond the frontier of sense perception. We observe man within a process of becoming. We do not simply observe mathematics on the one hand and sensory experience on the other but rather the emergence of mathematics within the developing human being. And now we come to that which truly leads over into spiritual science itself. You see, what we call forth out of our own inner life, this 'mathematical activity', becomes in the end an abstraction. Yet our experience of it need not remain an abstraction. In our time there is, to be sure, little opportunity for us to experience mathematics in a true light. Yet at a certain point in the development of Western civilization there does come to light something of this sense of a special spirit in mathematics. This comes to light at the point where Novalis, the poet Novalis, who underwent a good mathematical training in his studies, writes about mathematics in his *Fragments*. He calls mathematics a grand poem, a wonderful, grand poem.

One really must have experienced at some time what it is that leads from an abstract understanding of the geometrical forms to a sense of wonder at the harmony that underlies this inner mathematical activity. One really must have had the opportunity to get beyond the cold, sober performance of mathematics, which many people even hate. One must have struggled through as Novalis had in order to stand in awe of the inner harmony and — if I may use an expression you have heard often in a completely different context — the 'melody' of mathematics.

Then something new enters into one's experience of mathematics. There enters into mathematics, which otherwise remains purely intellectual and, metaphorically speaking, interests only the head, something that engages the entire human being. This something manifests itself in such youthful spirits as Novalis in the feeling that what you behold as mathematical harmony, what you weave through all the phenomena of the universe, is actually the same loom that *wove you* during the first years of growth as a child here on earth. This is to feel concretely man's connection with the cosmos. And when one works one's way through to such an inner experience, which many hold to be mere fantasy because they have not actually attained it themselves, one has some idea what the spiritual scientist experiences when he rises to a more extensive grasp of this inner mathematical activity by undergoing an inner development which you will find fully described in my book *How To Know Higher Worlds*.[62] For then the capacity of soul manifesting itself as this inner mathematics passes over into something far more comprehensive. It becomes something that remains just as exact as mathematical thought yet does not proceed solely from the intellect but from the whole human being.

On this path of constant inner work — an inner work far more demanding than that performed in the laboratory or observatory or any other scientific institution — one comes to know what it is that underlies mathematics, that underlies this simple faculty of the human soul which can be expanded into something far more comprehensive. In this higher experience of mathematics one comes to know Inspiration. One comes to understand the differences between what lives in us as mathematics and what lives in us as outer-directed empiricism. In this outer-directed empiricism we have sense impressions that give content to our empty concepts. In Inspiration we have something inwardly spiritual, the activity of which manifests itself already in mathematics, if we know how to grasp mathematics properly — something spiritual which in our early years lives and weaves within us. This activity continues. In doing mathematics we experience this in part. We come to realize that the faculty for performing mathematics rests upon Inspiration, and we can come to experience Inspiration itself by evolving into spiritual scientists. Our representations and concepts then receive their content in a way other than through external experience. We can inspire ourselves with the spiritual force that works within us during childhood. For what works within us during our childhood is spirit. The spirit, however, resides in the human body and must be perceived there through the body, within man. It can be viewed in its pure, free form if one acquires through the faculty of Inspiration the capacity not only to think in mathematical concepts but to view what exists as a real force, in that it organizes us through and through up until the seventh year. And what manifests itself partially in mathematics and reveals itself as a much more expansive realm through Inspiration can be inwardly viewed if one employs certain spiritual scientific methods. One thereby gains not merely new results to add to those acquired through the old powers of cognition but rather an entirely new mode of apprehension. One acquires a new 'inspirative' cognition.

## From projective geometry to Imagination

Perhaps for some of you it would be helpful to make an exact picture of how ordinary analytic geometry relates to so-called synthetic or projective geometry. I would like to say a few words on this subject. In analytic geometry we discuss some equation of the kind y = f(x). If we stay, for instance, in the x-y coordinate system, then we say that for every x there is a y, and we look for the points of the y-coordinate, which are the results of the equation. What is actually occurring here? Here we have to say that in the way we manipulate the equation, we always have our eye on something that lies outside of what we ultimately seek, because what we are really looking for is the curve.[63] But the curve or graph is not contained in the equation — only the possible x and y values are contained in the equation. When we proceed in this manner, we are actually working outside the curve; and what we get as values of the y-coordinate in relation to the x-coordinate we consider as points belonging to the curve. With our analytic equation, we never really enter the curve itself, its real geometric form. This fact has significant implication as regards human knowledge.

When we do analytic geometry, we perform operations which we subsequently look for spatially, but in all our figuring we actually remain outside of a direct con-templation of geometrical forms. It is important to grasp this, because when we consider projective geometry we arrive at a very different picture of what we are doing. Here, as most of you know, we don't calculate, we really only deal with the intersection of lines and the projection of forms. In this manner we get away from merely cal-culating around the geometrical forms, and we enter — at least to some degree — the geometrical forms themselves. This becomes evident, for example, when you see how projective geometry goes about proving that a straight line does not have two, but only one point at infinity. If we set off in a straight line in front of us, we will come back from behind us (this is easily understood from a geometrical point of view), and we can show that we travel through exactly one point at infinity on this line. Similarly, a plane has only one line at infinity, and the whole of three-dimensional space has only one plane at infinity.

These ideas, which I am only mentioning in passing, cannot be arrived at by analytical means. It is not possible. If we already have projective-geometric ideas, we may imagine we can do it; but we cannot really. However, projective geometry does show us that we can enter into the geometrical forms, which is not possible for analytic geometry. With projective geometry it is really possible. When we move out of mere analytic geometry into projective geometry, we get a sense of how the curve contains in itself the elements of bending, or rounding, which analytic geometry describes only externally. Thus we penetrate from the environment of the line, the surroundings of the spatial form, into its inner configuration. This gives us the possibility of taking a first step along the way from purely mathematical thinking—of which analytic geometry is the prime representative—to Imagination. To be sure, with projective geometry, we do not actually have Imagination yet, but we approach it. When we go through the processes inwardly, it is a tremendously important experience—an experience which can actually be decisive in leading us to an acknowledgment of the imaginative element. Also, this experience leads us to affirm the path of spiritual research inasmuch as we can form a real mental picture of what the imaginative element is.

## From phenomenon to Imagination

In ordinary waking life, you will agree, we are constantly perceiving, but actually in the very process of doing so we are continually saturating our percepts with concepts. In scientific thinking we interweave percepts and concepts entirely systematically, building up systems of concepts and so on. By having acquired the capacity for the kind of thinking that gradually emerges from *The Philosophy of Freedom*,[64] one can become capable of such acute inner activity that one can exclude and suppress conceptual thinking from the process of perception and surrender oneself to bare percepts. But there is something else we can do in order to strengthen the forces of the soul and absorb percepts unelaborated by concepts. One can, moreover, refrain from formulating the judgements that arise when these percepts are joined to concepts and create instead symbolic images, or images of another sort, alongside the images seen by the eye, heard by the ear, and rendered by the senses of warmth, touch, and so on. If we thus bring our activity of perception into a state of flux, infusing it with life and movement, not as we do when forming concepts but by elaborating perception symbolically or artistically, we will develop much sooner the power of allowing the percepts to permeate us as such. An excellent preparation for this kind of cognition is to school oneself rigorously in what I have characterized as phenomenalism, as elaboration of phenomena. If one has really striven not to allow inertia to carry one beyond the veil of sense perception upon reaching the boundary of the material world, in order to look for all kinds of metaphysical explanations in terms of atoms and molecules, but has instead used concepts to order the phenomena and follow them through to archetypal phenomena, one has already undergone a training that enables one to isolate the phenomena from everything conceptual. And if one also symbolizes the phenomena, turns them into images, one acquires a potent soul force enabling one to absorb the external world free from concepts.

Obviously we cannot expect to achieve this quickly. Spiritual research demands of us far more than research in a laboratory or observatory. It demands above all an intense effort of the individual will. If one has practised such an inner representation of symbolic images for a certain length of time and striven in addition to dwell contemplatively upon images that one keeps present in the soul in a way analogous to the mental representation of phenomena, images that otherwise only pass away when we race from sensation to sensation, from experience to experience; if one has accustomed oneself to dwell contemplatively for longer and longer periods of time upon an image that one has fully understood, that one has formed oneself or taken at somebody else's suggestion so that it cannot be a reminiscence, and if one repeats this process again and again, one strengthens one's inner soul forces and finally realizes that one experiences something of which one previously had no inkling. The only way to obtain even an approximate idea of such an experience, which takes place only in one's inner being — one must be very careful not to misunderstand this — is to recall particularly vivid dream-images. One must keep in mind, however, that dream-images are always reminiscences that can never be related directly to anything external and are thus a sort of reaction coming towards one out of one's own inner self. If one experiences to the full the images formed in the way described above, this is something entirely real, and one begins to understand that one is encountering within oneself the spiritual element that actuates the processes of growth, that is the power of growth. One realizes that one has entered into a part of one's human constitution, something within one; something that unites itself with one; something that is active within but that one previously had experienced only unconsciously. Experienced unconsciously in what way?

I have told you that from birth until the change of teeth a soul-spiritual entity is at work structuring the human being and that this then emancipates itself to an extent. Later, between the change of teeth and puberty, another such soul-spiritual entity, which dips down in a way into the physical body, awakens the erotic drives and much else as well. All this occurs unconsciously. If, however, we use fully consciously such measures of soul as I have described to observe this permeation of the physical organism by the soul-spiritual, one sees how such processes work within man and how man is actually given over to the external world continually, from birth onwards. Nowadays this giving-over of oneself to the external world is held to be nothing but abstract perception or abstract cognition. This is not so. We are surrounded by a world of colour, sound and warmth, and by all kinds of sense impressions. By elaborating these with our concepts we create yet further impressions that have an effect on us. By experiencing all this consciously we come to see that in the unconscious experience of colour and sound impressions that we have from childhood onwards there is something spiritual that suffuses our organization. And when, for example, we take up the sense of love between the change of teeth and puberty, this is not something originating in the physical body but rather something that the cosmos gives us through the colours, sounds and streaming warmth that reach us. Warmth is something other than warmth, light something other than light in the physical sense; sound is something other than physical sound. Through our sense impressions we are conscious only of what I would term external sound and external colour. And when we surrender ourselves to nature, we do not encounter the ether-waves, atoms, and so on of which modern physics and physiology dream; rather, it is spiritual forces that are at work, forces that fashion us between birth and death into what we are as human beings. Once we tread the path of knowledge I have described, we become aware that it is the external world that forms us. We become best able to observe consciously what lives and embodies itself within us when we acquire above all a clear sense that spirit is at work in the external world. It is of all things phenomenology that enables us to perceive how spirit works within the external world. It is through phenomenology, and not abstract metaphysics, that we attain knowledge of the spirit by consciously observing, by raising to consciousness what otherwise we would do unconsciously, by observing how, through the sense world, spiritual forces enter our being and work formatively upon it.

# Notes

1   Rudolf Steiner, *Anthroposophy and Science*, Lecture 3.
2   Rudolf Steiner identifies three forms of higher cognition: Imagination (in which spiritual realities are grasped in a form clothed in images); Inspiration (in which the spiritual reality behind the manifestations is grasped directly); and Intuition (in which the reality is recognized as a being with which one can unite one's will-forces). See, for example, *An Outline of Esoteric Science*, Chapter 5.
3   The age after the Atlantean civilization catastrophically ended, beginning some 10,000 years ago and extending into the modern age.
4   Nicholas Cusanus. 1401–64. He was a lawyer, cardinal, philosopher and mathematician, quoted by Steiner as an example of the consciousness of the time, which found itself increasingly unable to grasp the divine with emerging scientific and mathematical concepts.
5   The etheric body (also referred to as the life-body or formative-force body) is a body of life-forces which organizes and regulates the physical body. In addition, there is an astral body, which is the seat of consciousness and feeling, and an ego-body (or ego-organization), which is necessary for self-consciousness. See, for example, *An Outline of Esoteric Science*, Chapter 2, by R. Steiner.
6   Galen (AD 129–99) was the physician of the Roman emperor Marcus Aurelius. His numerous medical texts, based on his observations and dissections, provided the basis for

physiology and medicine for centuries to come. He was the first physician to use the pulse as a diagnostic aid.

7    Pneuma is later referred to as 'yellow gall', giving four fluids in all through which the external elements are related to the four members of the human being.

8    This statement comparing the laws of chemistry and physics may well be contentious today, but it is certainly true that nineteenth- and early twentieth-century chemical research revealed that, although many patterns exist in chemical combinations, there are also many exceptions. Chemistry is therefore more of an empirical science than physics, and indeed theoretical chemists strive to overcome its qualitative unpredictability by applying the concepts derived from pure physics.

9    See note 2.

10   Rudolf Steiner: *Riddles of Philosophy*, Anthroposophic Press, 1973.

11   Emil Du Bois-Reymond (1818–96) was a German physiologist who researched the electrical nature of nerve impulses. He was intensely interested in the problem of consciousness, and concluded that it was impossible to understand how physical processes could result in consciousness. His famous concept of 'Ignorabimus' (= we shall never know), delivered in a lecture to scientists and physicians in 1872, was a statement of the perceived limits of scientific enquiry. Similarly, he said we must assume the concept of 'matter' in the outer world, but we could never know its real essence.

12   Footnote by Rudolf Steiner: 'Herein lies the difference between organisms and machines. What is essential in the machine is only the interaction of its parts. The unifying principle that governs that interaction does not exist in the

object itself but outside it as a plan in the head of its builder. Only the most extreme short-sightedness can deny that the difference between an organism and a mechanism is precisely the fact that in a machine the determining principle governing the interrelationship of its parts is external (and abstract), whereas in an organism it assumes a real existence in the object itself. Thus, the sense-perceptible conditions of an organism do not appear merely to follow one from another, but are governed by an inner principle that is imperceptible to the senses. In this sense this principle is no more perceptible to the senses than the plan in the builder's head, which is also present only to the mind. Essentially, it is such a plan, except that it has entered the organism's inner being and affects it directly, not through a third party, the builder.'

13  That is, from beyond the realm of sense-perceptible nature.

14  A footnote by John Barnes, editor of *Nature's Open Secret* from which this text is extracted, stresses the importance of this new scientific method: 'The distinction he makes here between *concept* and *idea* has important consequences, particularly for the methodology of organic science. The dynamic, living idea of the *type* leads research along a very different path and to a very different result than do the analytical methods of inorganic science. Grasping the idea of the *type* requires a different quality of thinking (here called *reason* or, in the following, *judgement through intuitive perception*) which is essential to Goethe's method of organic science. It is only this kind of active, participatory thinking that can apprehend the type in its various metamorphoses. Thus, exact, participatory thinking becomes the method of organic science.'

15   *Italian Journey*, 17 May 1787, included in the 'Report' from July 1787.

16   Footnote by Rudolf Steiner: 'The fruit develops through the growth of the lower part of the pistil [the ovary, *i*]; it represents a later stage of the pistil and can therefore only be drawn separately. The formation of the fruit is the final expansion of the plant. Its life now differentiates itself in an organ that closes itself off from its environment—the fruit and the seed. In the fruit everything has become appearance; it is outward appearance only, and estranges itself from life to become a dead product. All the essential inner life impulses of the plant are concentrated in the seed, from which a new plant arises. The seed has become almost entirely idea; its outer appearance has been reduced to a minimum.'

17   *Italian Journey*, 2 December 1786.

18   From: *The Metamorphosis of Animals*, by Goethe.

19   Ibid.

20   Footnote by Rudolf Steiner: 'In modern science, the term *primal organism* usually refers to a primal cell (primal cytode), a simple entity at the lowest stage of organic evolution. This is a very specific, concrete, sense-perceptible entity. In a Goethean sense, the term *primal*, or *archetypal*, *organism* does not refer to this but to the essence (being), or formative entelechy principle, that makes the "primal cell" into an organism. This principle manifests in both the simplest and the most perfected organisms, but developed differently. It is the animality in the animal, and that which makes a living being into an organism. Darwin assumes it from the beginning; it is there, is introduced, and then he says that it reacts one way or another to environmental

influences. For Darwin, it was an indefinite X; Goethe wanted to explain this indefinite X.'

21  The hypothetical 'ether' of physics is not to be confused with the ether (and etheric forces) spoken of by Steiner. The former was largely discredited in Steiner's lifetime (though there have been some attempts to re-evaluate it), whereas the latter is perceptible to spiritual cognition.

22  Centimetre per second per second is the accepted unit for acceleration, in contrast to velocity, which is just centimetre per second.

23  Lines drawn at 90 degrees to a surface, which feature in many ray-diagrams.

24  See note 5 above.

25  Fresnel (1788–1827) conducted experiments (some of which had been carried out earlier by Young) in which light from two sources, projected onto a screen, produces a latticework of light and dark (an 'interference pattern'). This dealt a severe blow to the corpuscular theory of light, and led to the widespread adoption of a wave theory of light.

26  In other words, the effects of these planets can be traced in a watery medium.

27  Presumably birds.

28  In Lecture 1 of the 'Warmth Course', Rudolf Steiner disputes the usual theories of the sun's structure, based on earthly concepts. He makes the remarkable assertion that in the sun 'there is less than empty space'. This concept of negative space or counter-space has proved fruitful in understanding the ethers. See 'Further Reading', p. 237.

29  This refers to the fact that all gases, irrespective of their chemical nature, react to heat in approximately the same way. The gas laws relating pressure, temperature and

volume, apply equally (to a good degree of accuracy) to all gases. Hence the phrase 'unifying influence of the sun'.

30    The Accademia del Cimento was a learned society in Florence, founded in 1657 and disbanded ten years later. They wished to test experimentally some of the tenets of Aristotle's natural philosophy. They also developed the work of Galileo on the expansion of liquids when heated, producing a thermometer that was an improvement on Galileo's original invention. According to some authorities, the Academy marked the beginning of modern physics, a view not shared by Steiner, although he does point to the strong materialistic influence it exerted.

31    This reference is to imaginative cognition, see note 2.

32    In other words, the liquid prevents the minute portions on the surface from falling through the remainder of the liquid.

33    Meaning the way particles are assembled in a gas.

34    Presumably in order to breathe.

35    See references under 'Further Reading', pp. 237–8.

36    Tone (or sound) is usually explained as the ear's response to waves of rarefaction and condensation in the air. Rudolf Steiner insists that tone has an independent reality, so that rarefaction and condensation merely *accompany* the tone, that is, the waves are *carriers* of tone, and it is wrong to identify them as *being* the tone.

37    According to orthodox physics with its wave theory, light which passes through certain substances (tourmaline crystals in this case) is said to be *polarized*; it vibrates in one plane only. If such light is then passed through a second polarizing crystal, the amount of light which gets through depends on the mutual orientation of the two crystals. If their axes are

aligned, all the light passes through, whereas if the crystals are crossed at right angles to each other, no light gets through. Furthermore, if other solids such as small crystals are placed between the two crossed polarizing crystals, and viewed, every slight variation of density is visible as beautiful colour fringes. This can easily be reproduced at home by taking the two lenses from a pair of polaroid sunglasses and placing them together, with a plastic ruler sandwiched in between. The stress lines in the plastic can then be seen. According to Rudolf Steiner, such phenomena arise through the direct working of one form on another, the light simply making this working visible.

38   A glucoside contained in horse-chestnut bark.

39   The science of photochemistry studies the effects of light on chemical reactions. It is generally accepted that ultraviolet light shows the most chemical effects, and even in everyday life it is well known that the ultraviolet rays of the sun can cause changes (which may be undesirable) in the skin. Curiously, in photosynthesis, it is the *red* light which has the effect.

40   See Chapter 8. Imagine dropping lots of little stones; the way they fall under gravity to produce a *surface* gives a picture of a liquid.

41   The term 'pressure' is here used loosely to mean forces which act in space on physical objects; and 'suction' is the rather difficult concept of the opposite of this, described as 'negative gravity' in Chapter 9.

42   During the 'Warmth Course' Rudolf Steiner performed a number of experiments, including the interconversion of energy. In the experiment mentioned here, steam in a piston was condensed to water, causing the piston to move, i.e.,

heat energy was converted into work. He also demonstrated the reverse, i.e., the conversion of work into heat.

43   In modern physics this is explained by including the concept of entropy, i.e., some of the energy must be expended in changing the degree of disorder of particles (the 'entropy' of the particles), so that amount of energy is not available to do mechanical work.

44   Goethe described colour as arising from the 'deeds and sufferings of light'.

45   In Lecture 12 of the 'Warmth Course', Rudolf Steiner developed mathematical ideas which could be used to systematize the various realms of nature. In particular, he suggested the use of negative numbers to describe chemical effects, imaginary numbers (involving $\sqrt{-1}$), to describe light effects, and super-imaginary numbers in dealing with the z realm, which he characterizes as life effects.

46   Mass here means the same as 'body' or 'entity'.

47   The concept of 'relief' is here used to state that light experiments stand apart, stand out like the picture on the wall, independently of their surroundings.

48   See next chapter for more about the life ether.

49   The term 'ether' is introduced in Lecture 12 of the 'Warmth Course' (not reproduced here). The essential being of warmth is called *warmth ether*. Similarly, the x, y and z realms are designated *light ether*, *chemical ether*, and *life ether*.

50   A concept in physics referring to the eventual even distribution of heat throughout the cosmos: as heat leaves the hotter bodies and becomes evenly distributed, so there is no more flow of energy. This final (hypothetical) state would not allow any further change to occur, so all life and evolution would cease.

51   In the final lecture of the 'Warmth Course', Steiner speaks of acoustics, linking it to the y realm of chemical effects. Tone arises when the y realm works in the gaseous (x') realm. Hence, 'chemical ether' is also known as 'tone ether'.

52   Presumably Steiner means non-living building materials, e.g. stone etc.

53   In the photoelectric effect, electricity is produced when light falls on certain metals.

54   See *Outline of Esoteric Science*, Chapter 6.

55   The world of pure spiritual archetypes comprises, broadly speaking, Lower Devachan and Higher Devachan. See Steiner's *Theosophy of the Rosicrucian*, Lecture 8. Each of these planes has its negative (or evil) reflection, referred to here. The same lecture course also contains descriptions of the three classes of adversary forces: Lucifer, Ahriman and the Asuras.

56   There have been many speculations about this 'third force' of fallen life ether. It may well be connected with the forces unleashed by atomic physics, though it is likely that there is more to it than that.

57   The 'spiritual soul', or consciousness soul as it is also called, bestows the faculty of objective understanding. It is the member of the soul which has unfolded since the fifteenth century, and which is responsible, amongst other things, for the modern scientific approach to nature.

58   See *The Temple Legend*, Lecture 10 (p. 123), Rudolf Steiner Press, London 1985.

59   Democritus, the Greek philosopher of about 470–380 BC is generally credited as being the father of the atomic theory.

60   Compare Chapter 1 of his book *How to Know Higher Worlds*, where Rudolf Steiner gives a meditative exercise in which

the attention is directed to budding, growing, flourishing life on the one hand, and fading, decaying, withering phenomena on the other hand.

61   See *An Outline of Esoteric Science*, Chapter 4, for a description of these earlier phases in the evolution of the Earth.

62   See 'Further Reading', p. 237.

63   The equation can be plotted as a curve or graph.

64   Steiner recommended working through this book as an exercise towards developing a thinking that is independent of the senses, a thinking capable of understanding the non-material. It cannot be read merely as a source of information. In Lecture 8 of *The Boundaries of Natural Science*, he says, 'In a sense, the book is only a kind of musical score that one must read with inner thought activity in order to progress, as the result of one's own efforts, from one thought to the next. The book constantly presupposes the mental collaboration of the reader.'

# Sources

1 Condensed from Lecture 2, *Origin of Natural Science*, Anthroposophic Press, 1985.

2 Condensed from Lecture 3 of the above volume.
'Ptolemy, Copernicus and Newton' are from Lecture 4 of the above volume, pp. 44–5, 47–8, 50.

3 Lecture 7 of above volume, pp. 86–95, and Lecture 8, p. 103.
'From ritual to experiment' is from *Anthroposophy and Science*, Mercury Press, 1991, pp. 88–92.

4 *The Boundaries of Natural Science*, Anthroposophic Press, 1983, pp. 9–14.
'Where does thinking go wrong' is taken from pp. 21–4 of the above volume.
'Goethe's use of thinking' is from *Anthroposophy and Science*, pp. 98–100.

5 Condensed from Chapter 4 of *Nature's Open Secret*, Anthroposophic Press, 2000, pp. 42–5, 49–52, 55–63.

6 *Light Course*, Steiner Schools Fellowship Publications, 1987, pp. 2–11.

7 *Light Course*, pp. 18–23.
'The primary phenomenon of colour' is from *Light Course*, pp. 35–8, 43–4.

8 'What did the Greeks mean?' is from *Warmth Course*, Mercury Press, 1988, pp. 27–30.
'Solid, liquid and gas between earth and cosmos' is from *Warmth Course*, pp. 66–70.

'First steps of a new method' is from *Warmth Course*, pp. 75–9.

9   'Not just energy' is from *Warmth Course*, pp. 81–2.

'Negative gravity and negative form' is from *Warmth Course*, pp. 83–8.

'Warmth between the physical and the spiritual' is from *Warmth Course*, pp. 101–5.

10   *Warmth Course*, pp. 109–12.

'Spatial, non-spatial, pressure and suction' is from *Warmth Course*, all of Lecture 11 (pp. 131–42).

11   *Warmth Course*, pp. 156–63.

'The ethers in the human organism' is from the 'Bridge Lecture' in *Course for Young Doctors*, Mercury Press, 1994, pp. 229–30.

12   *Etherization of the Blood*, Rudolf Steiner Press, 1971, pp. 28–9, 39–42.

'From nature to sub-nature' is from *Anthroposophical Leading Thoughts*, Rudolf Steiner Press, 1973, pp. 216–19.

13   'What did Democritus mean by "atoms"?' is from *Origins of Natural Science*, pp. 23–4.

'Atomism versus continuism' is from *Origins of Natural Science*, pp. 63–7.

'Spiritual conception of atoms' is from *The Karma of Materialism*, Anthroposophic Press, 1985, pp. 33–4.

14   'Scientific concepts in social life' is from *The Social Future*, Anthroposophic Press, 1972, pp. 8–10.

'Science and the individual' is from *The Karma of Materialism*, Anthroposophic Press, 1985, pp. 31–2.

'Natural science as a foundation for spiritual science' is from *The Boundaries of Natural Science*, pp. 32–3.

'Mathematics: origins and future possibilities' is from *The Boundaries of Natural Science*, pp. 36–40.

'From projective geometry to Imagination' is from *Anthroposophy and Science*, pp. 84–6.

'From phenomenon to Imagination' is from *The Boundaries of Natural Science*, pp. 109–112.

# Further Reading

## By Rudolf Steiner

*Origins of Natural Science*, Anthroposophic Press, 1985
*Anthroposophy and Science*, Mercury Press, 1991
*The Boundaries of Natural Science*, Anthroposophic Press, 1983
*Light Course*, Anthroposophic Press, 2003
*Warmth Course*, Mercury Press, 1988
*Etherisation of the Blood*, Rudolf Steiner Press, 1971
*Nature's Open Secret*, Anthroposophic Press, 2000
*Astronomy Course* (out of print but available on Internet:
    www.awakenings.com/astronomy)
*The Philosophy of Freedom*, Rudolf Steiner Press, 4th imp., 1999
*An Outline of Esoteric Science*, Anthroposophic Press, 1997

## By other authors

Bockemühl, J. (ed.), *Toward a Phenomenology of the Etheric World*,
    Anthroposophic Press, 1985
Bortoft, Henri, *Goethe's Scientific Consciousness*, Institute for Cul-
    tural Research, 2nd ed., London, 1998
Edwards, Lawrence, *The Field of Form*, Floris Books, 1982
Edwards, Lawrence, *The Vortex of Life*, Floris Books, 1993
Jones, Michael, *Nuclear Energy: a Spiritual Perspective*, Floris
    Books, 1983
Lehrs, Ernst, *Man or Matter*, Rudolf Steiner Press, 1985
Lehrs, Ernst, *Spiritual Science, Electricity, and Michael Faraday*,
    Rudolf Steiner Press, 1975

Marti, Ernst, *The Four Ethers*, Schaumberg, 1984

Merry, Eleanor, *Goethe's Approach to Colour*, extracts edited by John Fletcher, Mercury Arts Publications, 1987 (Rudolf Steiner Press)

Thomas, Nick, *Science Between Space and Counterspace*, New Science/Temple Lodge, London, 1999

Unger, Georg, *On Nuclear Energy and the Occult Atom*, Anthroposophic Press, 1982

Whicher, Olive, *Sunspace*, Rudolf Steiner Press, 1989

# Note Regarding Rudolf Steiner's Lectures

The lectures and addresses contained in this volume have been translated from the German, which is based on stenographic and other recorded texts that were in most cases never seen or revised by the lecturer. Hence, due to human errors in hearing and transcription, they may contain mistakes and faulty passages. Every effort has been made to ensure that this is not the case. Some of the lectures were given to audiences more familiar with anthroposophy; these are the so-called 'private' or 'members' lectures. Other lectures, like the written works, were intended for the general public. The difference between these, as Rudolf Steiner indicates in his *Autobiography*, is twofold. On the one hand, the members' lectures take for granted a background in and commitment to anthroposophy; in the public lectures this was not the case. At the same time, the members' lectures address the concerns and dilemmas of the members, while the public work speaks directly out of Steiner's own understanding of universal needs. Nevertheless, as Rudolf Steiner stresses: 'Nothing was ever said that was not solely the result of my direct experience of the growing content of anthroposophy. There was never any question of concessions to the prejudices and preferences of the members. Whoever reads these privately printed lectures can take them to represent anthroposophy in the fullest sense. Thus it was possible without hesitation — when the complaints in this direction became too persistent — to depart from the custom of circulating this material "For members only". But it must be borne in mind that faulty passages do occur in these reports not revised by myself.' Earlier in the same chapter, he states: 'Had I been able to correct them [the private lectures], the restriction *for members only* would have been unnecessary from the beginning.'

The original German editions on which this text is based were published by Rudolf Steiner Verlag, Dornach, Switzerland in the collected edition (*Gesamtausgabe*, 'GA') of Rudolf Steiner's work. All publications are edited by the Rudolf Steiner Nachlassverwaltung (estate), which wholly owns both Rudolf Steiner Verlag and the Rudolf Steiner Archive. The organization relies solely on donations to continue its activity.

For further information please contact:

Rudolf Steiner Archiv
Postfach 135
CH-4143 Dornach

or:

www.rudolf-steiner.com

Rudolf Steiner

# AGRICULTURE
An Introductory Reader

Compiled with an introduction, commentary and notes by
Richard Thornton Smith

The evolving human being
Cosmos as the source of life
Plants and the living earth
Farms and the realms of nature
Bringing the chemical elements to life
Soil and the world of spirit
Supporting and regulating life processes
Spirits of the elements
Nutrition and vitality
Responsibility for the future

ISBN 1 85584 113 4

Rudolf Steiner

# ARCHITECTURE
An Introductory Reader

Compiled with an introduction, commentary and notes by
Andrew Beard

The origins and nature of architecture
The formative influence of architectural forms
The history of architecture in the light of mankind's spiritual
    evolution
A new architecture as a means of uniting with spiritual forces
Art and architecture as manifestations of spiritual realities
Metamorphosis in architecture
Aspects of a new architecture
Rudolf Steiner on the first Goetheanum building
The second Goetheanum building
The architecture of a community in Dornach
The temple is the human being
The restoration of the lost temple

ISBN 1 85584 123 1

Rudolf Steiner

# ART
An Introductory Reader

Compiled with an introduction, commentary and notes by
Anne Stockton

The being of the arts
Goethe as the founder of a new science of aesthetics
Technology and art
At the turn of each new millennium
The task of modern art and architecture
The living walls
The glass windows
Colour on the walls
Form—moving the circle
The seven planetary capitals of the first Goetheanum
The model and the statue 'The Representative of Man'
Colour and faces
Physiognomies

ISBN 1 85584 138 X

Rudolf Steiner

# EDUCATION
An Introductory Reader

Compiled with an introduction, commentary and notes by
Christopher Clouder

A social basis for education
The spirit of the Waldorf school
Educational methods based on anthroposophy
The child at play
Teaching from a foundation of spiritual insight and education in
    the light of spiritual science
The adolescent after the fourteenth year
Science, art, religion and morality
The spiritual grounds of education
The role of caring in education
The roots of education and the kingdom of childhood
Address at a parents' evening
Education in the wider social context

ISBN 1 85584 118 5

Rudolf Steiner

# MEDICINE
An Introductory Reader

Compiled with an introduction, commentary and notes by
Dr Andrew Maendl

Understanding man's true nature as a basis for medical practice
The science of knowing
The mission of reverence
The four temperaments
The bridge between universal spirituality and the physical
The constellation of the supersensible bodies
The invisible human within us: the pathology underlying
therapy
Cancer and mistletoe, and aspects of psychiatry
Case history questions: diagnosis and therapy
Anthroposophical medicine in practice: three case histories

ISBN 1 85584 133 9

Rudolf Steiner

# RELIGION
An Introductory Reader

Compiled with an introduction, commentary and notes by
Andrew Welburn

Mysticism and beyond: the importance of prayer
The meaning of sin and grace
Rediscovering the Bible
What is true communion?
Rediscovering the festivals and the life of the earth
Finding one's destiny: walking with Christ
The significance of religion in life and death
Christ's second coming: the truth for our time
Universal religion: the meaning of love

ISBN 1 85584 128 2